Lessons Learned Troubleshooting Steam Heat

Ray Wohlfarth

This book does not take the place of the boiler manufacturers written instructions, engineering, or code issues that may be in force in your locale. Please follow the boiler manufacturer's instructions included in the boiler installation manual. This does not take the place of a properly designed system from an experienced designer. Thank you for choosing to purchase and read this book.

DEDICATION

This is dedicated to my amazing family: Sheila, Jon, Conor & Lyndsay, Abby & Mike & their children, Delaney & Owen, Samantha & Ryan & their children, Jake & Annie. I also must acknowledge my friend who started my on this path, Dan Holohan. Thanks for the gentle nudge. Thank you, Dave Smoyer, for your expertise and input.

Table of Contents

Boiler Room Safety

A boiler room or basement can be a dangerous place. Be sure you are working in a safe environment. The following are suggestions I have for a safe service call.

Plan your escape – If the room fills with steam, it will have no visibility and steam displaces oxygen. Look around for any trip hazards if you must leave in a hurry.

Shut off boilers until you are sure the room is safe.

Is there carbon monoxide in the room? I suggest you use a personal CO detector when entering.

Is there combustion air? Is it large enough? Is it blocked?

Do you smell natural gas or sour flue gases. This should be corrected right away.

Is the flue for the water heater and boiler intact, properly pitched, and safe?

Is the boiler or water heater jacket discolored or has black streaks? This could indicate carbon monoxide forming.

If your boiler room was flooded and the boiler or burner was under water, do NOT operate the boiler until it is checked by an expert. Consult the boiler manufacturer as they most likely will tell you to replace the boiler.

Is there rust under the water heater or boiler draft diverter? This could indicate back drafting.

Is there anything dangerous stored in the boiler room? Gasoline, lawn mower, or snow blower. The following a list compiled by Weil McLain of things to avoid storing in a boiler room.

10 most common steam boiler room problems we see:
1. The near boiler piping is wrong or too small.
2. Insufficient combustion air.
3. Boiler pressure set too high.
4. Pipes are leaking.
5. Insufficient space around boiler.
6. Lack of boiler maintenance.
7. Main air vent in wrong location.
8. Missing pipe insulation
9. Pop safety valve piped incorrectly.
10. Boiler gauge glass is flooded.

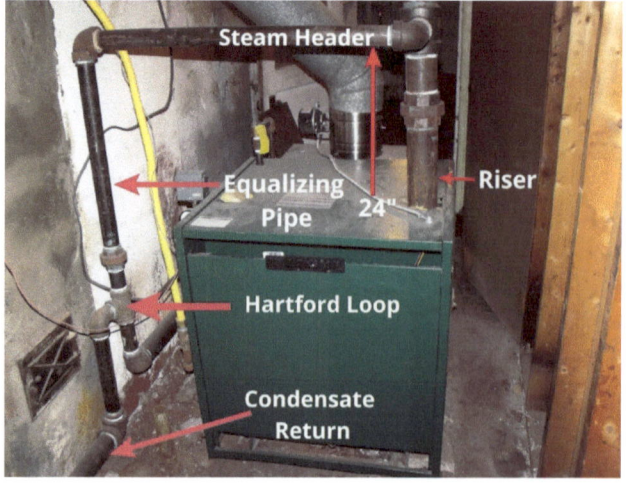

Boiler Room Detective

I have always equated steam system troubleshooting to being like being a detective. You must evaluate the clues you see, hear, or smell and come to a theory. The most important thing you will need when diagnosing steam problems is patience, sometimes lots of patience. You make an adjustment, cycle the burner numerous times, and walk away feeling proud of yourself. A day, a week, a month later; the problem returns. I explain this to the customer and inform them I will start with the least expensive repairs before moving on to the more expensive solutions.

Rarely is it one thing causing the problem you are seeing. Instead, it is usually something in one area causing the reaction where you are. Understanding steam systems requires a more holistic approach. For example, I was called about a corner classroom that heated sporadically. By the time I arrived, the maintenance staff replaced the steam trap twice and the shutoff valve once. We followed the condensate piping and discovered the solution in the crawl space below the building. A clevis hanger came apart and the pipe sagged, creating a water seal. This seal prevented air from leaving or entering the piping for the radiator. After lifting the pipe and reattaching the pipe hanger, the system worked great.

While servicing steam systems may sound complicated, I believe you know how to troubleshoot steam systems, you just didn't know you knew. The following are suggestions I have for troubleshooting steam systems:

Shelve all your knowledge about troubleshooting air or hydronic systems, it will mislead you when servicing steam systems. Steam works completely opposite of other hvac systems. For example, if you want more air from a blower, you could increase the fan speed. To get more heat from a hydronic system, you could increase the water temperature. If you increase the pressure in a steam system, it will have the opposite reaction. You will get less steam to the system. I know, weird right?

Steam's only goal is to surrender its latent heat and revert back to water. It will look for the closest place to do that. If the horizontal steam main feeding the building has condensate due to missing pipe insulation, the steam will condense there and you will have no heat at the ends of the system, no matter how many times you replace the trap or air vent.

Think like steam is a way to help troubleshoot steam systems. For example, a 16-ounce bottle of water, if converted to steam, would fill one hundred feet of 2-inch pipe. It requires one Btu to raise a pound of water one degree Fahrenheit until the water temperature reaches 212^0F. To convert 212^0F water at atmospheric pressure to steam, requires an additional 970 Btus. This is called latent heat. While that seems extreme, the latent heat energy is not wasted. It zooms through the piping looking for somewhere cool to surrender the heat and become water.

My YouTube channel, Boiler Room Detective, shows videos on troubleshooting steam and hydronic heating systems. Good luck Detective!

Dissolved oxygen is 10 times more corrosive than CO2

Steam Rules

Troubleshooting steam heating systems goes against your instincts as it is completely opposite of air or hydronic systems. These are five rules I follow for steam heating:

Rule #1 Steam systems need to breathe.

Rule #2 Pressure goes from high to low.

Rule #3 Two gases cannot occupy the same space.

Rule #4 The problem and the solution are rarely in the same room.

Rule #5 Always assume the boiler was installed wrong until you decide otherwise.

Rule #1 Steam systems need to breathe –

When the system is off, all the pipes and above the water line in the boiler are filled with air. As the steam starts, it will push the air through the pipes and out the vent. When the steam stops, it will condense, and air will rush in to take the place of the steam.

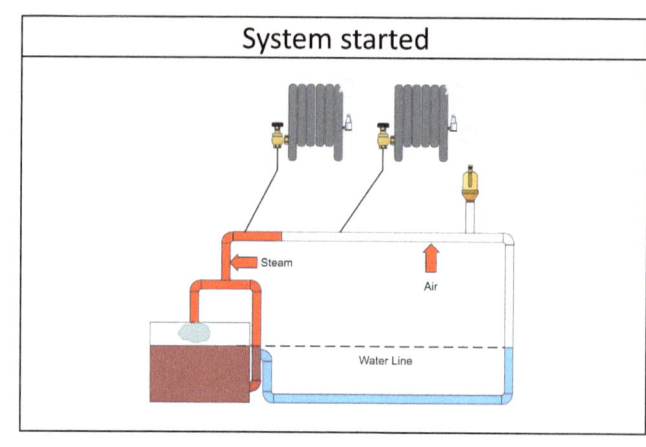

System started

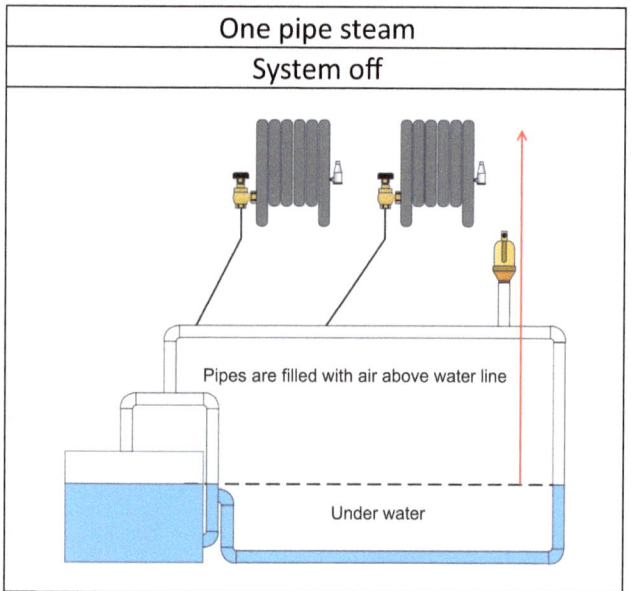

One pipe steam

System off

Pipes are filled with air above water line

Under water

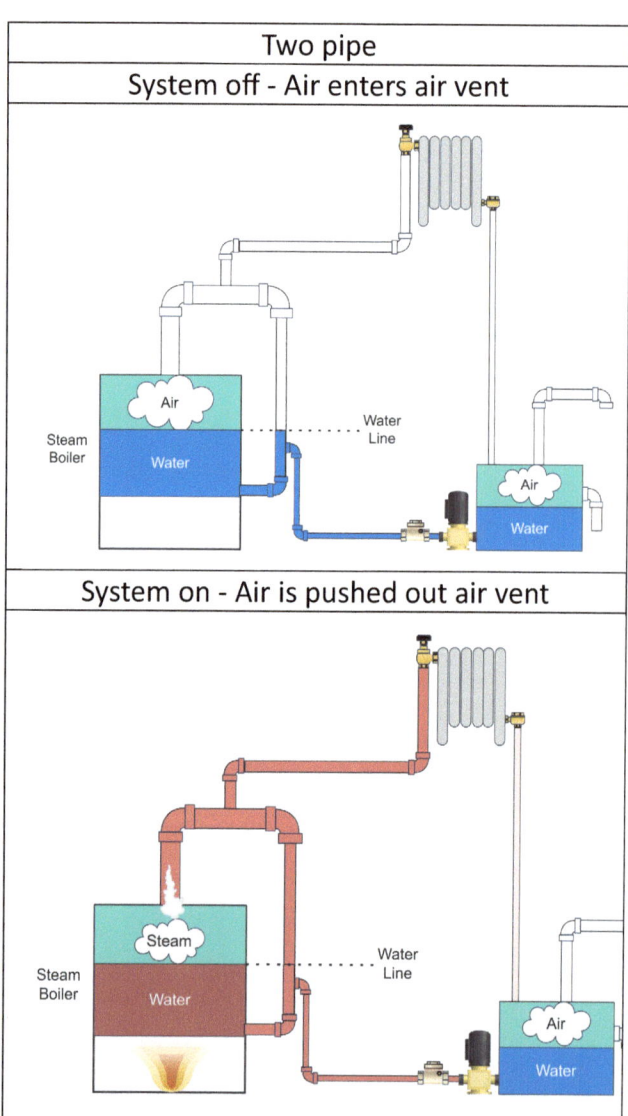

Two pipe

System off - Air enters air vent

System on - Air is pushed out air vent

Water expands @ 1,600 its volume when it becomes steam.

Rule #2 Pressure goes from high to low - Steam doesn't care about your flow arrows. It will travel in any direction from higher to lower pressure. If steam condenses upstream, this lowers the pressure, and the steam may go backward.

Rule #3 Two gases cannot occupy the same space - If the radiator or pipe is filled with air, steam cannot enter. Once the air is removed, the steam can enter. At the end of the heating cycle, the steam condenses, and air will rush in to take the place of the steam.

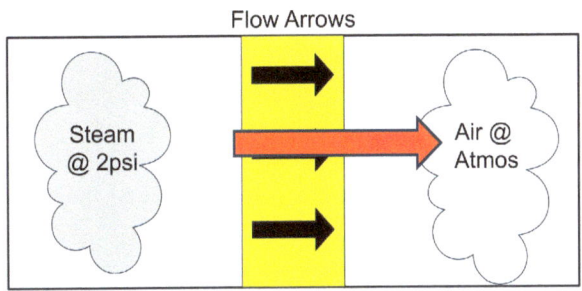

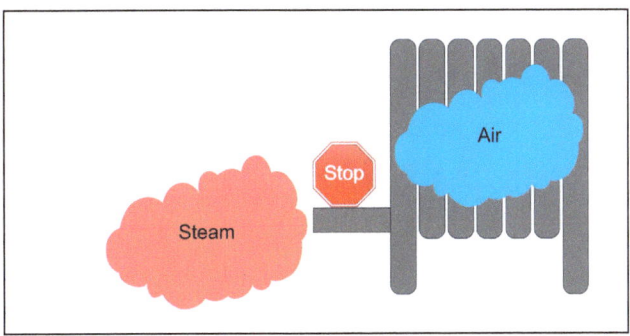

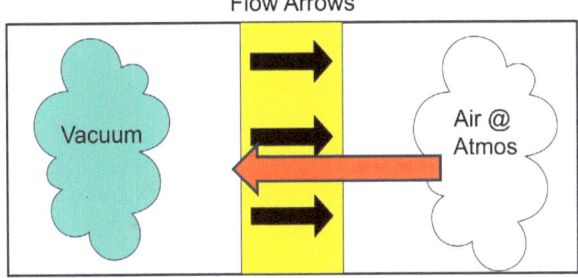

A compound gauge is a valuable service tool. It helps you know if boiler is going into a vacuum.

One pipe radiator

Steam @ 2psi has six times the volume of steam @ 100 psi.

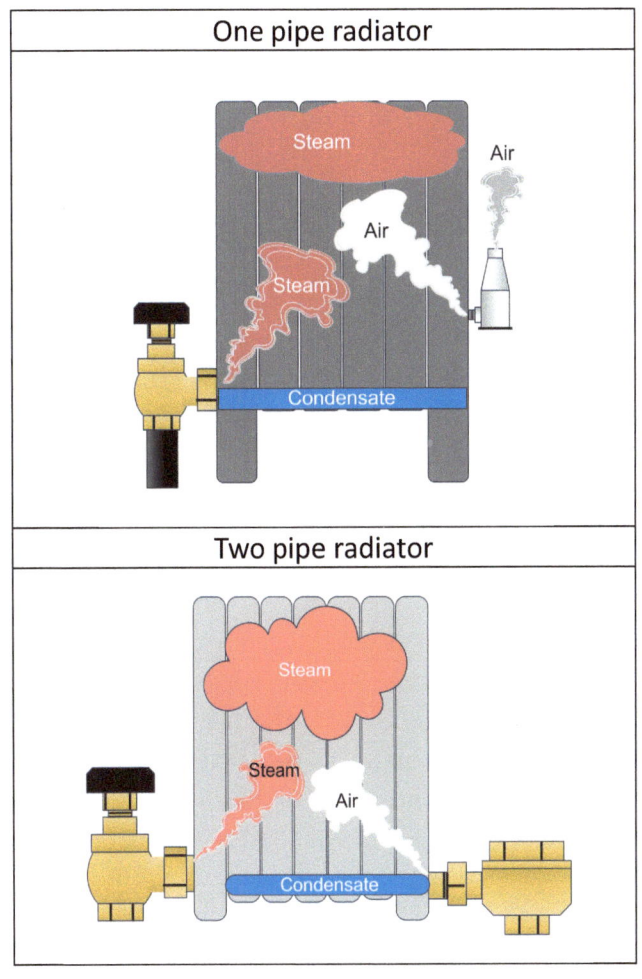

Two pipe radiator

Rule #4 The problem and the solution are rarely in the same room -When working on steam systems, a leaking steam trap in another room may cause the radiator in the problem room not to heat. We had a two-pipe radiator not heating in one room and the cause of the problem was a sagging pipe two stories below on the condensate return pipe.

Rule #5 Always assume the boiler was installed wrong until you decide otherwise - Most boilers are installed by the low bidder who didn't know about steam. Once you touch the system, you own it and all the problems.

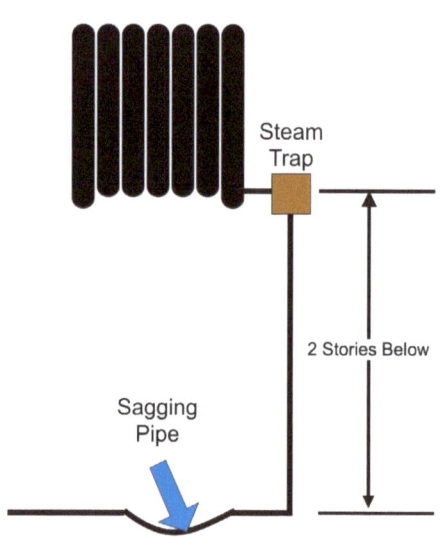

Incorrect location of air vent	Correct location of air vent
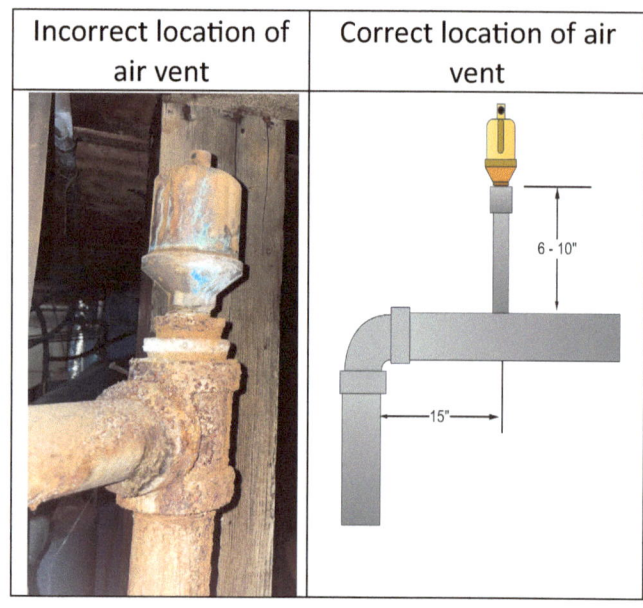	

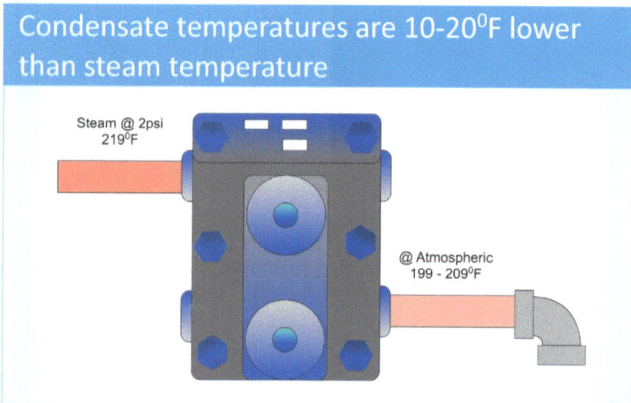

The leading cause of boiler accidents is improper maintenance.

Types of Heat	
Sensible Heat	Sensible heat is any heat transfer that causes a change in temperature without causing a change of state. Sensible heat can be measured with a dry bulb thermometer.
Latent Heat	Latent heat is the amount of heat required to cause a change of state. In a boiler system, this would be the amount of heat added to water to change it from water to steam. It requires 970 Btus to raise 1 pound of water at 212^0F to steam.
Total Heat	Total heat is the sum of the sensible and latent heat in an exchange process. It is sometimes called enthalpy.

Boiler Problems

The boiler is not firing.	
Power	If there is no power, check the breaker, switch, manual rest limit control, or the emergency door switch.
Call for heat	Is there a call for heat?
Steam pressure	Check the pressure gauge to see if the boiler is off on steam pressure.
Fuel	Verify the gas pressure is at the inlet to gas train. If not, look for closed gas valves.
Flame failure	If the boiler is off due on flame failure, press the reset button Once. If it trips again, have the burner checked to determine the cause. *Do not press reset more than once without checking the burner!*
Low water cutoff	Verify water is present in gauge glass. Press the manual reset on the auxiliary low water cutoff.
Gas pressure switch	Gas pressure switches on the gas train assure the gas pressure to the boiler is correct. These controls have a manual reset button.
Limit control	Verify boiler steam pressure is lower than the steam pressure limit control setpoint. The boiler limit control is a manual reset control.
Pilot	Verify pilot solenoid valve opens and pilot gas pressure regulator has proper gas pressure. Most flame failures occur during the pilot sequence.

Emergency door switch	The door switch located just in/outside the boiler room could be pushed or switched off.
Adjustable bleed valve	The adjustable bleed valve, used to slow the opening of the gas valve, could be closed. This is located on the gas valve. Open valve a half turn and try again.
Gas train vent	The vent could be plugged, stopping the gas valve from opening. Try disconnecting the vent tubing from valve & restart.

Water under boiler	
Check boiler for leaks	Shut off the boiler and look at the fireside to see if you see any leaks.
Check piping for leaks	Look around boiler room for pipe leaks. Repair the leaks ASAP.
Cold water in boiler	When the boiler is starting with room temperature water, water could drip. If the leak continues after the boiler starts to steam, look for a boiler leak.

Lag boiler is flooding	
No high-level spill trap	ASME code recommends using high level spill traps on each boiler to prevent flooding.

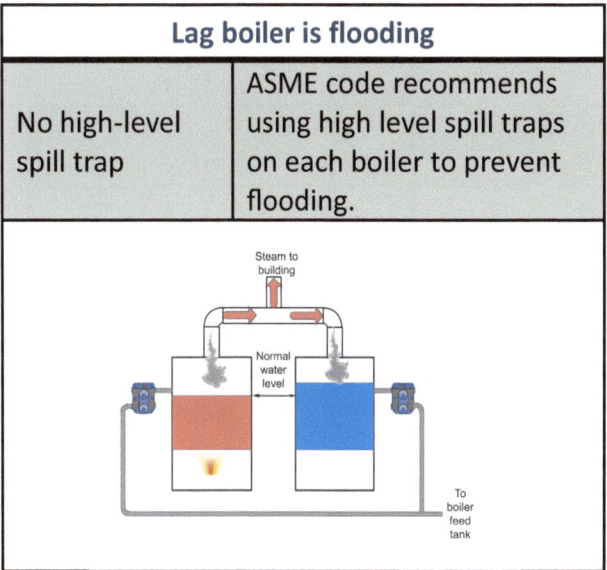

Intermittent Flame Failures	
Loose wire	Verify each wire terminal screw is snug.
Electrical voltage	Verify electrical voltage is within range.
Pilot solenoid	Verify pilot solenoid valve opens.
Pilot gas pressure regulator	Verify pilot pressure doesn't drop during pilot sequence.
Enable relay is a Triac type relay	Triac relays allow voltage leaks through before relay contacts fully make. Replace with definite purpose relay.
Variable frequency drive or VFD Drive is too close	Install a power line "noise" filter at the supply side of the boiler's power-on switch.
Draft proving switch	Verify draft proving switch is working properly. Check for blockage in draft sensing tube.
Gas pressure	Verify gas pressure does not drop below minimum setting. If there is more than one appliance in the room, check gas pressure when all are operating.
Flame safeguard	A failing flame safeguard control could cause the flame failures.
Pilot flame	A pilot flame can be pulled away from the flame sensor.
Dirty flame sensor	Try cleaning the flame sensor.
Defective flame detector	The ultraviolet or infrared flame detector could be weak and failing.
Defective gas valve actuator	Look for leaking hydraulic fluid in window of valve actuator. If it is, replace the actuator.

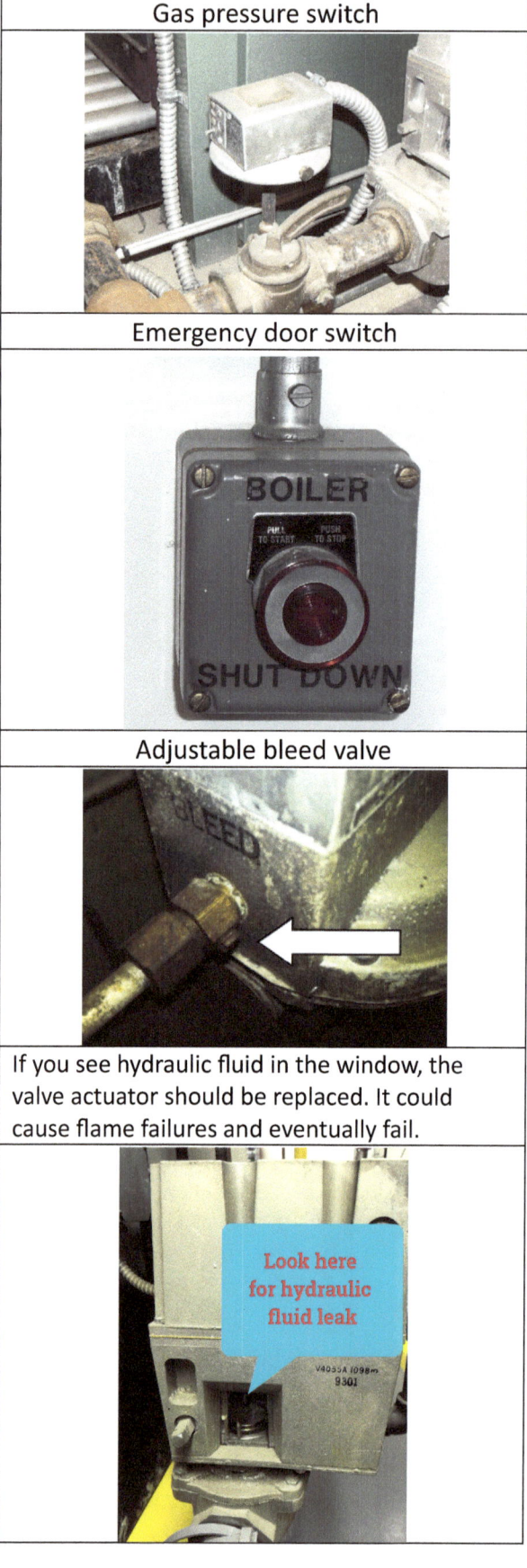

Gas pressure switch

Emergency door switch

BOILER

SHUT DOWN

Adjustable bleed valve

If you see hydraulic fluid in the window, the valve actuator should be replaced. It could cause flame failures and eventually fail.

Look here for hydraulic fluid leak

Burner short cycles	
Flame rod is dirty or misaligned	The flame rod senses the pilot flame. Check flame signal & clean or replace the flame rod.
Flame detector is defective	Check flame detector for proper flame signal.
Steam piping is air bound	If the air cannot leave the piping, the steam cannot enter. Look for closed valves or sags in the piping.
Pressure differential too close	The differential on the steam pressure control could be too close. Widen the differential.
Chimney draft too high	The draft may be pulling the flame away from the flame sensor.
High ph or TDS in boiler water	If the pH or TDs is high, it could cause the boiler to trip on low water, resulting in the burner short cycling.
Defective pressure control	Verify pressure control works properly.
Boiler losing call for heat	Verify the boiler has a call for heat. It could be a thermostat losing the call or another safety control.

Rust under boiler or water heater draft diverter	
Negative conditions	If you see rust under the boiler or the draft diverter, it could be due to a negative condition in the boiler room
Underfired burner	If the burner is not firing to capacity, the flue gases could be condensing.
Blocked flue	Check for flue blockage
Blocked fireside of boiler	Verify the fireside is not plugged with soot or debris.

High fuel costs	
Undersized or Underfired boiler	Undersized boiler will run excessively
Burner short cycling	If the burner starts and stops, it uses more energy.
Someone installed 1 pipe air vent on a 2-pipe radiator	The air vent causes the steam to change directions resulting in slow return & should never be done.
Air vent not working	Air is trapped inside piping and steam cannot get in.
Incorrect near boiler piping	Carryover could form due to incorrect near boiler piping.

Rust under draft diverter

Boiler draft is wrong
Boiler draft is too low
Excessive heat in boiler
Flue gas spillage into room
Could allow Carbon Monoxide to form
Affects boiler efficiency
Possible damage to burner
Flame impingement on burner head
Boiler draft is too high
Elevated flue temperatures
Lowers boiler efficiency
Could allow Carbon Monoxide to form
Flame impingement on metal surfaces

Discoloration of boiler jacket	
Flue gas leak	If the hot flue gases are not going up the flue, they could cause discoloration of the boiler jacket.
Exhaust fan	An exhaust fan in the boiler room could pull the flue gases from the boiler and into the room.
Insufficient combustion air	If the combustion air opening is blocked or restricted, it could cause the flue gases to roll out of the boiler and into the boiler room.
Plugged flue passages	If the flue passages are plugged, the flue gases could spill from the boiler into the boiler room.
Refractory	Refractory could be missing or damaged
Combustion air	The combustion air may have changed. Look for closed or blocked combustion air openings.

Discoloration of jacket could signify dangerous back drafting of flue gases and boiler should not operate until checked.

Discolored boiler jacket

Steam can travel at 60 MPH through pipes.

Flame rolls from under boiler base (Atmospheric type boiler)	
Negative condition	Exhaust fans in the building could pull the boiler room into a negative. -0.03" WC could pull flue gases from boiler.
Insufficient combustion air	Verify sufficient combustion air
Blocked flue	Check for flue blockage
Blocked fireside of boiler	Verify the fireside is not plugged with soot or debris. Look for boiler leak.
Gas pressure regulator downstream of the electric valve?	Verify gas pressure regulator is upstream of electric gas valves.
Valve opening too fast	Adjust bleed orifice in gas valve

Flame rollout

Boiler water is priming	
Overfiring the boiler	Check burner firing rate
Excess chimney draft	Verify draft is not excessive.
Fluctuations in steam demand	Check for quick opening steam valves or vacuums formed by the closed valves.

Boiler water is foaming	
Elevated solids in boiler	Blowdown boiler to lower the solids in the water
High pH in boiler water	Elevated pH can cause foaming & carryover.

The boiler is sooted	
Negative condition in boiler room	Boiler room could be in negative condition due to exhaust fans in building.
Insufficient combustion air	Verify the room has enough combustion air.
Blocked flue	Check for flue blockage
Blocked fireside of boiler	Verify the fireside is not plugged with soot or debris.
Incorrect air to fuel adjustment	Check the efficiency of the boiler with an analyzer. Adjust as needed. Verify gas pressure is not too high.
Boiler is leaking	Steam or water leaking from the boiler could cause sooting.
Soot is very dangerous and extreme care required when cleaning it. See the sooted boiler below.	

The flue gas temperature is higher than before	
Scale	Scale could be forming on the water side of the boiler. The water treatment specialist should be notified. Scale impedes the heat transfer in the boiler.
Dirty fireside	If the fireside of the boiler is dirty, this could result in higher temperatures and fuel consumption.
Overfired	The burner may be overfiring in the boiler, causing more heat than the boiler can handle.
Excess draft	The draft or velocity of the flue gases traveling through the boiler is too high. Results in insufficient heat transferred to the boiler.

How soot affects boiler efficiency

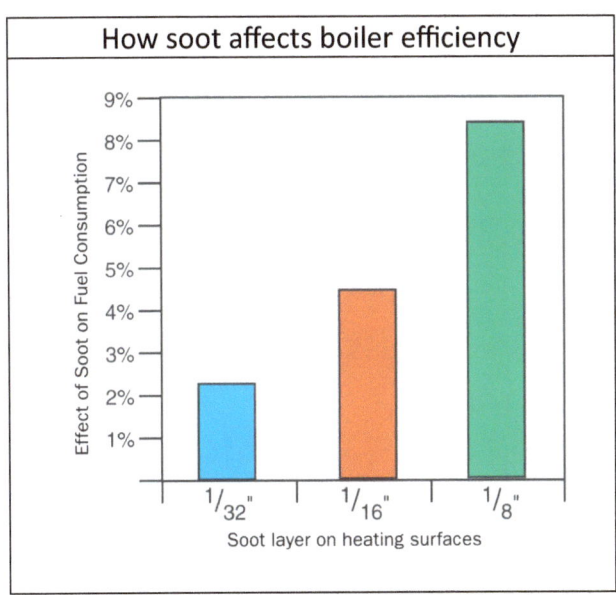

Boiler going into a vacuum	
Sagging pipe	A sagging pipe could allow water to pool, cause premature condensation of the steam, & form a vacuum. Remove places where water can pool in the piping.
Missing pipe insulation	Uninsulated pipes could cause premature condensation of the steam & form a vacuum. Replace the missing insulation.
Closed valve	A closed hand or automatic valve could stop air from reentering system.
System can't breathe	Steam systems must breathe. View the piping to be sure air can leave and return.

The boiler is flooding	
Low water cutoff	The low water cutoff may require service or could be defective.
System vacuum	The system could be going into a vacuum and may need a vacuum breaker.
Combination low water cutoff/feeder is defective	A plugged or defective control can continue filling the boiler.
Boiler feed pump	The boiler feed pump contactor may be hanging up and keeping the pump operating.
Feedwater valve leaking	The feedwater valve for the boiler may be leaking through.
No high-level spill trap	If there are two or more steam boilers in a building, each boiler should have a high-level spill trap to avoid flooding of the idle boiler.
Water quality	The boiler water may be dirty and cause the boiler feed unit to overfeed the water.

Boiler can't build steam pressure	
Undersized or underfired boiler	Verify boiler is large enough to handle the steam load
Boiler is leaking	Check for steam or water leaks. A leaking boiler is very dangerous.
Carryover	Carryover can cause premature condensation of steam. Check near boiler piping.
Steam is condensing too quickly	Verify steam piping is insulated and water cannot pool in the piping. Check the pitch of the piping.
Pressure gauge span	A low-pressure boiler typically has a 30-psi gauge. If someone replaced it with one with a larger scale, it could look like it's not building pressure
Pigtail is plugged	If the pigtail is plugged, the pressure gauge may not read the pressure.
Pressure gauge is defective	If the gauge is defective, it may not read the pressure.
Air vent is panting	A panting air vent is usually caused by water pooling & condensing the steam.

Boiler losing water	
System leak	Find & repair any system water leaks.
Boiler leak	If boiler is leaking, it may have to be replaced.
Steam pressure too high	If the steam pressure is too high, the water can be lost when the condensate flashes to steam and exits via the vent.
Leaking steam traps	If a steam trap is blowing through, steam could be lost through the tank vent.

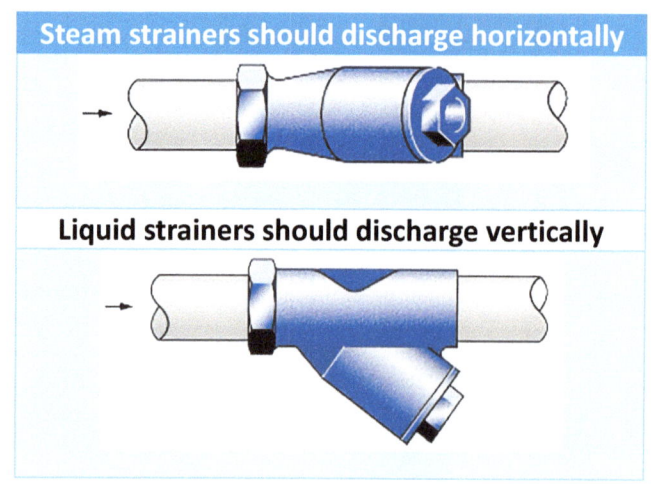

Steam strainers should discharge horizontally

Liquid strainers should discharge vertically

Pop safety valve opening	
Plugged pigtail	If the pigtail is plugged, the pressure controls may not read the boiler steam pressure and continue firing.
Defective pressure control	If the pressure control is defective, it may shut not shut off the burner at the setpoint.
Defective pop safety valve	If the pop safety opens often, the internal spring could weaken and open at a lower pressure.
Pressure controls set too high	If the steam pressure controls are set too close to the pop safety valve relieving pressure, the valve could open.
Isolation valve on boiler	When the steam boiler shuts off, the hot surfaces continue to make steam. If a valve on the boiler outlet closes right after the burner stops, the pressure inside the boiler could continue to rise.

Drip pan ell is used to keep the weight of the discharge piping off the pop safety valve.

Pop safety discharge piping

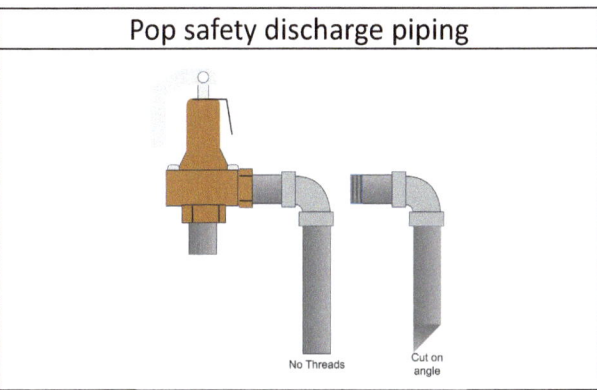

When testing a pop safety valve, most manufacturers want it tested when the boiler pressure is close to the safety valve setpoint. I like testing it when the steam pressure is above 11 psi.

A thermal imaging camera can help you find hidden steam pipes or problems behind the wall

The Titanic ship had 29 boilers & 129 furnaces.

Gauge Glass

Is boiler gauge glass empty or flooded?

You can tell by holding a pencil behind the gauge glass. If pencil appears intact, (Top picture) the gauge glass is empty. If pencil looks broken, (Lower picture) the gauge glass is full.

Gauge glass is flooding	
Low water cutoff	The low water cutoff may be defective and calling for water. Check the float or control.
System is in vacuum	The system may need a vacuum breaker.
Makeup water valve leaking through	Check makeup water valve for system. Dirt may be on the seat allowing water to leak through.
Steam pipe pitched wrong	If the steam pipe is pitched back to the boiler, condensate could flow back to boiler when it's off.
Restricted / plugged wet return	If the wet return is restricted with dirt and rust, the system may be introducing fresh water before the water returns.
No high-level spill trap	If there are two or more boilers, each should each have a high-level spill trap, about 2 inches above boiler water line.
Leaking check valve on feedwater pump pipe	Water from the boiler feed tank could be leaking into the boiler if tank is higher than boiler water level. Use spring loaded check valve instead of swing check.

Water level in gauge glass drops below glass	
Surging or priming in boiler	This could cause the water level to bounce up and down. Check TDS.
Carryover	Steam could be pulling water into piping.
Valve opening in the system	If a steam valve suddenly opens, it could cause erratic water levels in the boiler.
Boiler does not have enough capacity	If the boiler is underfired or undersized, it could pull the water out of the boiler.
Feedwater not working	Verify water can get into the system. Check makeup water valve or boiler feed operation.
Boiler is leaking	Check for leaks in boiler.

Broken gauge glass?

If gauge glass breaks, close bottom valve first and then the top valve. Always ensure the vertical rods are in place as they are there to protect the gauge glass from breaking.

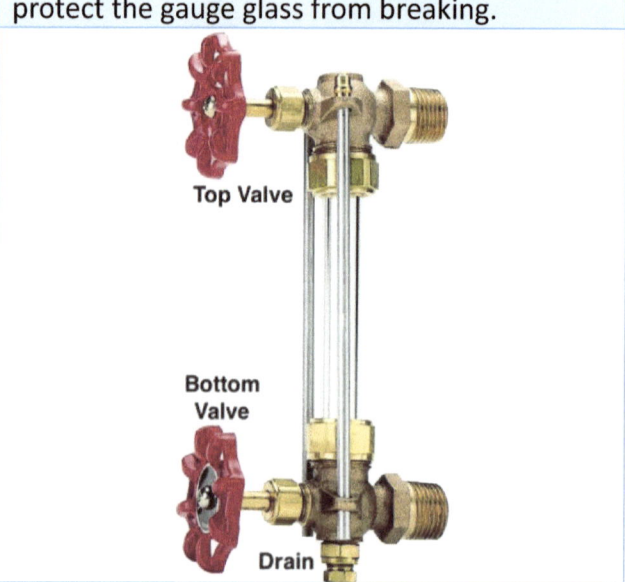

Water is bouncing in the gauge glass	
Oil in the boiler	Oil from pipe threading or from the manufacturing process could cause surging and bouncing.
Test the boiler TDS	Elevated TDS or Total Dissolved Solids in the boiler water could allow foaming / priming and carryover, which can cause wildly bouncing water levels.
Control valve opening too quickly	If a steam valve opens too quickly, the water can be pulled from the boiler and cause it to trip.
Chemical treatment	The water treatment chemicals may be overfed to the boiler causing bouncing.
Boiler too small	If the boiler is too small, it could cause bouncing water levels.
High chimney draft	Excess draft can cause burner to overfire
Burner overfired	If the burner is over fired, the water could bounce.
Vacuum in system	If there are automatic radiator valves with no vacuum breaker, it could cause the vacuum.
Internal boiler restrictions	If the boiler water side is filled with mud, it could cause bouncing water. Clean & flush the boiler.

Gauge glass is leaking steam or water	
Nut is loose on the gauge glass	Try tightening the nut slightly. Be prepared to replace the gauge glass.
Top of gauge glass is very thin.	Check the boiler water pH. High pH can cause steam etching eroding the top of the gauge glass.

Blowdown Gauge Glass

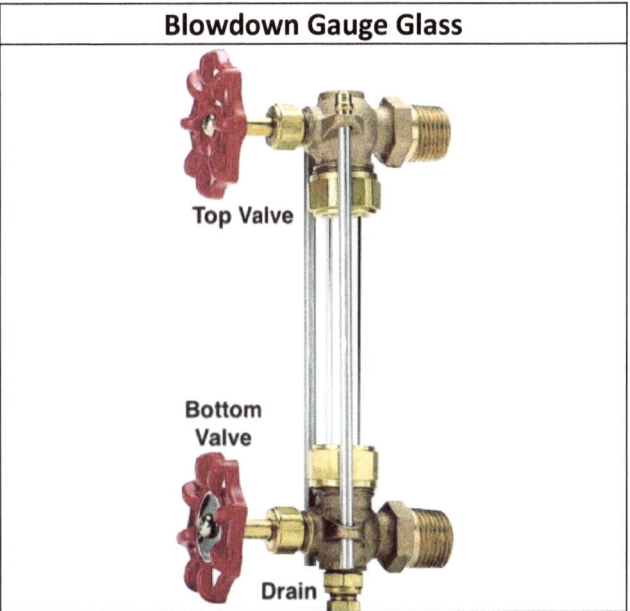

Top Valve

Bottom Valve

Drain

Use safety glasses and protective clothing for this. The discharge of the drain will be very hot!
Close Top and Bottom valves.
Open Drain valve.
Open Top valve and let it blow for a few seconds and then close top valve. If no steam/water blows out, the pipe connection into the boiler could be plugged & should be cleared before operating the boiler.

Open Bottom valve and let water blow for a few seconds. If no steam/water blows out, the pipe connection into the boiler could be plugged and should be cleared before operating the boiler.
Close the bottom valve.
Close the Drain valve.
Open the Bottom valve.
Open the Top Valve.
The water level should be about an inch above the cast mark in the McDonnell Miller #150 low water cutoff. If the gauge glass is thin or brittle, replace it.

The average life of properly treated condensate pipe is 10-20 years.

Types of low water cutoffs

Float type low water cutoffs

Probe low water cutoffs

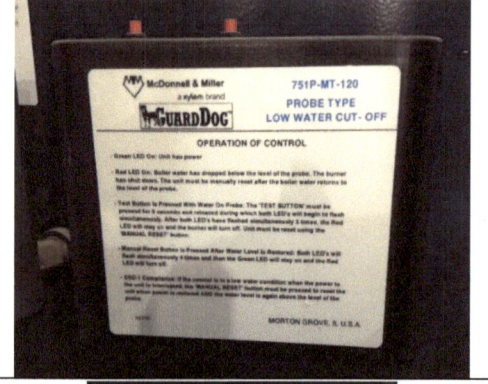

Boiler going off on low water

Condensate pump not running	Verify pump is running at the proper pressure, typically 20 psi.
Test the boiler TDS or Total Dissolved Solids	Elevated TDS in the boiler water could cause foaming, priming, and carryover.
Oil on the water	Pipe threading oil can find its way to the top of the water & cause rocking of the boiler water. Oil floats on top of the water in the gauge glass. You could try draining and flushing the system, but you may need to clean it, depending on the amount of oil in the water.
Control valve opening too quickly	If a zone valve opens too quickly, the water can be pulled from the boiler.
System going into a vacuum	Vacuums can form in the system if the air is restricted from entering when the steam condenses. Look for places where water can pool.
Low water cutoff control is defective	The combination pump control / low water cutoff could be filled with dirt or mud.

Flash steam percentage at various steam pressures

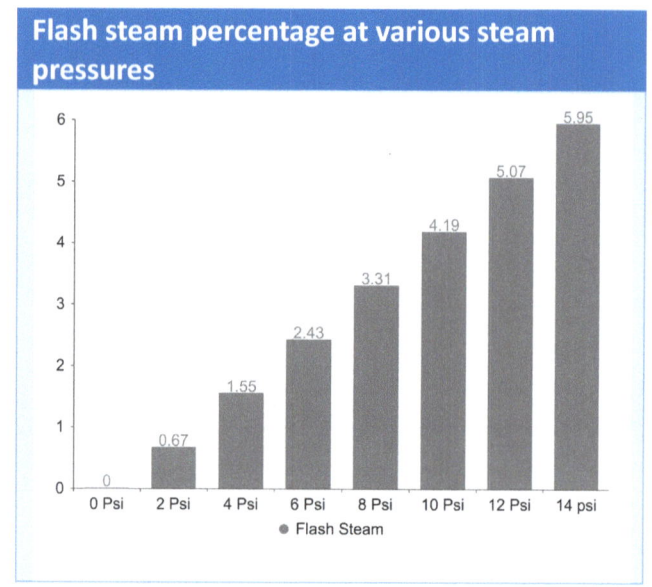

Steam System Water Hammer Possible Cause

Water Hammer @ Startup
Sagging pipes could cause condensate to pool
Steam main drip traps Verify they are working & have enough static head above.
"A" dimension is too short
Verify dry return is not a wet return
Steam piped into bull of tee
Check zone valve piping Verify water doesn't pool against steam valve when closed
Steam header not pitched properly
Concentric reducers in horizontal steam piping

Water Hammer @ End of Heating Cycle
Hartford Loop connection too close to the boiler water line
Pipe insulation removed in boiler room.

Water Hammer During Mid Heating Cycle
Clogged returns could cause water to back up into header
Excess boiler draft
Verify near boiler piping is correct
Boiler oversized / overfired
Defective / Leaking steam traps
Water seal in front of condensate tank
One pipe radiator venting too quickly. Try using two smaller ones
Master trap in piping. This could allow water to pool.
Failed steam trap stays closed
Excessive flow from condensate pump could shoot into the equalizing pipe. Verify flow is correct and equalizer is large enough.
Long nipple in Hartford Loop It should be a close nipple
Pipe insulation removed will cause premature steam condensation.
Return pipe not pitched properly. The pipe must pitch toward the condensate return tank.
Pipe sag before condensate or boiler feed tank
Zone valve is not trapped Be sure water cannot accumulate against zone valve seat.

How radiator covers affect heat output

The heat output from a cast iron radiator is 60% convection & 40% radiation. When you cover a steam radiator, it prevents convection & the heat output is reduced.

5% loss of capacity

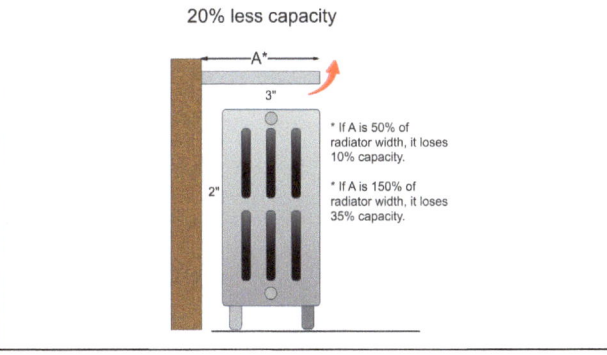

20% less capacity

* If A is 50% of radiator width, it loses 10% capacity.

* If A is 150% of radiator width, it loses 35% capacity.

30% loss of capacity

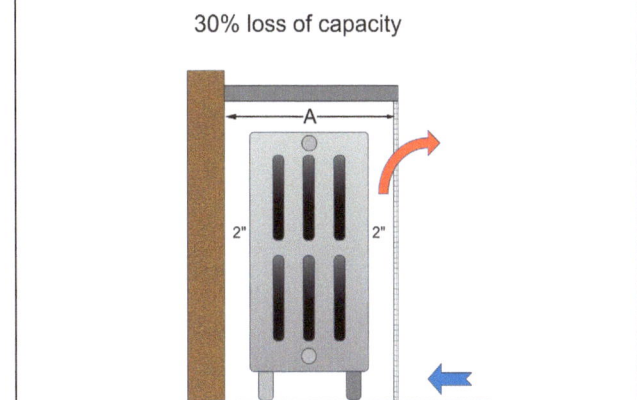

Steam pressure controls

Manual reset limit pressure control trips	
Operating control set too high	Operating pressure control is set too close to the manual reset limit control. Try adjusting limit pressure control higher.
Operating control is defective	If the operating control is set for a lower pressure, and the burner keeps firing until it reaches the setting of the limit pressure control, the operating control is defective.
Pigtail / siphon is plugged	If the pigtail is plugged, the control cannot sense the boiler pressure. Replace the pigtail. Always use a brass pigtail.
Isolation valve closed	When the burner shuts off, the internal surfaces of the boiler continue to make steam. Is the boiler steam valve closed? The pressure could rise until it reaches the limit control setting.
Limit control set too low	The manual reset control may be set too close to the operating pressure control.
Limit control is defective	Watch the steam pressure as the boiler fires. Note where the limit control shuts off the burner. If it shuts off before the pressure setting, replace the control.

Steam pressure controls are either additive or subtractive. On additive controls, set the pressure on the front to where the boiler starts. The differential is how much higher above the setpoint the control shuts off the burner. On subtractive controls, set the setpoint to the pressure you want the boiler to stop. The differential is how many pounds the boiler drops until it restarts.

Manual reset limit control I like setting this for 10 psi when the system is operating at 2 psi.

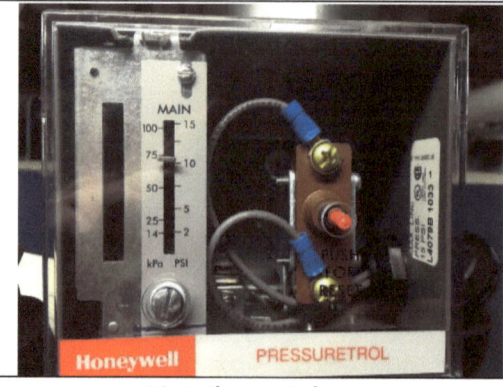

Pigtails or siphons

Water seal instead of pigtail

Never mount all the controls on one pigtail. It could plug and disable all the pressuretrols.

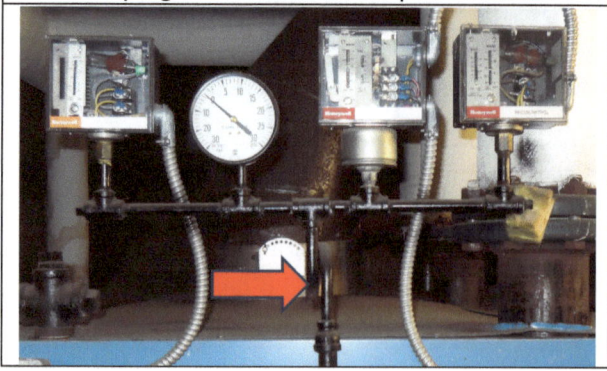

Steam Pressure Operating Controls
Additive control
This setting is the starting pressure of the boiler, typically set for 0.5 psi.

This is the differential pressure setting, typically set for 1 ½ psi. The boiler will operate between ½ and 2 psi when there is a call for heat.

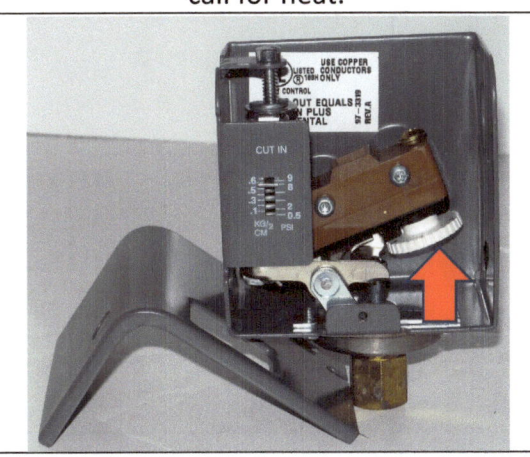

Did you know vinegar lowers pH and baking soda raises pH?

Subtractive control
This setting is the pressure where you want the boiler to shut off, typically 2 psi.

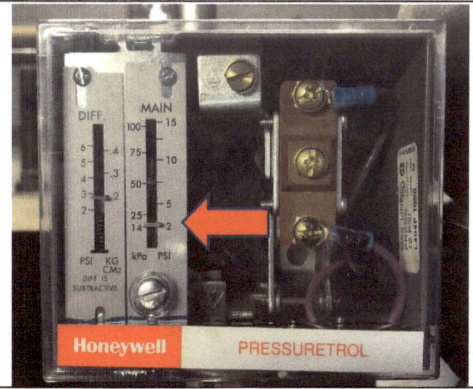

This is the differential pressure where the boiler will restart when there is a call for heat, typically ½ psi. You have to guesstimate where it is.

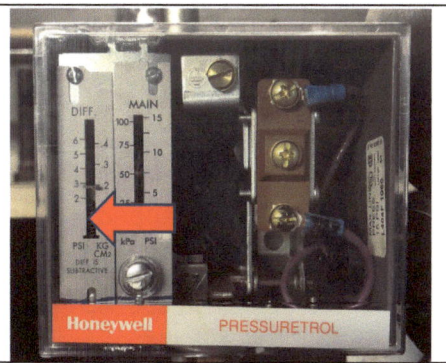

Firing Rate or Modulating Control
This is typically set for a steam pressure just below the operating pressure.

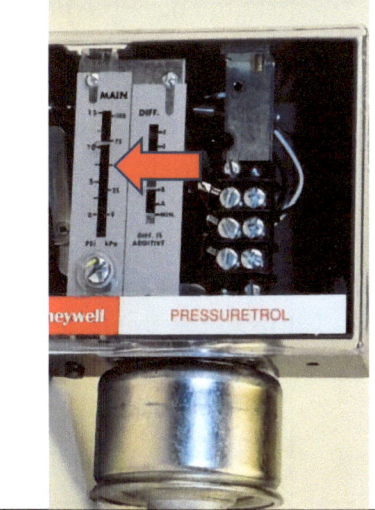

Boiler noises

Pipes are banging	
The system was flooded	Flooding could cause water to pool in the piping and get picked up by the steam.
Installer used a concentric reducer instead of eccentric one	The concentric reducer allows water to pool in the piping while an eccentric reducer will not.
Steam traps not working	If a steam trap is defective, allowing steam to leak into the condensate side of the system, this could meet the condensate from the other traps and cause water hammer which will ruin the nearby traps.
Steam pipes not properly dripped	The steam piping should be dripped every 150 feet to remove condensate which may have formed. It should also be dripped whenever the steam rises or at the end of the steam main.
Drip legs too small	The drip legs should be the same pipe size as the steam main it is dripping up to four-inch pipe size.
Strainer is full of water	The strainer before a trap should have the drain horizontal and not vertical.
Sagging pipes	Look for a sagging pipe on either the steam or condensate side. Water can pool and be picked up by the steam.
Missing pipe insulation	If the insulation was removed from the steam side of the system, this could cause the steam to condense too rapidly and allow the water to pool
Quick closing valve	The steam velocity is suddenly stopped when the valve closes, and all that momentum could cause the condensate to slam into the valve.
The boiler Is overfired	The overfired boiler could cause carryover.
No drip leg before control valve	If the condensate pools in front of a closed valve, it will be picked up by the steam and hurled into the closest fitting.
Pipe pitch is incorrect	The piping should be pitched so condensate cannot pool. Verify the piping is pitched, usually 1 inch every 10 feet.
Near boiler piping is incorrect	The near boiler piping is used to dry the steam before entering the piping. If the piping is wrong, it could allow water to leave the boiler, which could cause banging.
Boiler is surging or priming	This could cause carryover and the banging pipes. Check the pH, TDS, and chemical levels in the boiler water.

Squealing noise when boiler runs	
Defective blower / inducer	The motor bearings may be defective.
Blower wheel	If the blower wheel is mis adjusted, it could squeak from rubbing on housing.

The boiler makes a moaning sound when firing	
Flexible gas pipe	Sometimes gas flow through the flexible csst tubing is noisy. Verify it is properly sized.

The boiler makes a ticking /cracking noise when firing

Scale	Scale could be forming on the water side of the boiler. The water treatment specialist should be consulted.
Low Water	The boiler water level may be too low. If so, do not add water to the boiler until the boiler is shut off and cool.
Sludge inside boiler	Try draining the sludge from the boiler. You may need to hose the sludge from inside the boiler.

The boiler makes a humming or buzzing sound

Bound motor	Look for a bound electrical motor such as the burner blower, induced draft fan or feedwater pump.
Loose / broken wire	Check all wiring connections and snug the screws on the terminal. Look for arcing.
Defective relay	Check the condition of the relays. Replace the defective ones.
Wrong voltage	Verify the electrical voltage is within the proper range.

The boiler makes a hissing sound when firing

Boiler is leaking	Boiler water can be dripping onto burner flame. There could be a steam leak above the boiler water line.

Excessive combustion noise

Improper air to fuel adjustment	Verify air to fuel ratio is correct
High gas pressure	Verify the gas pressure feeding the burner is correct and not overfiring.
Excess draft	Check the draft and verify it's not too high.
Fast opening gas valve	If the gas valve opens too quickly, it could cause rumbling. Try closing the adjustable orifice on gas valve vent.
Improper flue pipe	Verify the flue pipe is sized properly and has correct pitch.
Blocked flue passage	Verify the flue is free and clear and not blocked.

Excessive vibration in system

Defective burner / feedwater pump motor	The bearings may be going in the burner motor. Replace the motor if defective.
Improper combustion	Check the air to fuel ratio
Closed valve on feedwater system	Verify the feedwater pump discharge pipe is not plugged or restricted. Look for closed valves.
Blower wheel	If the blower wheel is dirty or some of the balancing weights are missing, it could cause vibration. Replace blower wheel before it ruins blower motor bearings.

Did you know old timers used to place potatoes inside the boiler to control scale?

Water Treatment

Boiler pH is typically 7-9		
pH Readings		
PH Readings	Compare to	Acid/ Alkaline
0	Battery Acid	ACID
1	Hydrochloric Acid	ACID
2	Lemon Juice, Vinegar	ACID
3	Grapefruit /Orange Juice	ACID
4	Acid Rain /Tomato Juice	ACID
5	Black Coffee	ACID
6	Urine / Saliva	ACID
7	Neutral	
8	Sea Water	ALKALINE
9	Baking Soda	ALKALINE
10	Milk of Magnesia	ALKALINE
11	Ammonia	ALKALINE
12	Soapy Water	ALKALINE
13	Bleach / Oven Cleaner	ALKALINE
14	Drain Cleaner	ALKALINE

Carryover causes	
Overfired burner	Verify burner firing rate is correct.
Excess solids	Check TDS in boiler water. Blow down boiler if high.
Fluctuating load	If the steam demand varies quickly, it could cause carryover.
Excess draft	Excess draft could cause the boiler to overfire. Adjust draft control.
Oil in piping	Oil forms a film on top of water causing bouncing water levels.
Overfeeding chemical treatment	If the water treatment chemicals are high, it could cause elevated TDS.
Steam pressure too low	If boiler steam pressure is set too low, the steam velocity is increased causing carryover.
Incorrect near boiler piping	This could cause carryover if piping is wrong or undersized.
Steam zone valve	Improperly installed zone valve could cause carryover. Verify there is a vacuum breaker.

Excessive chemical treatment use	
Chemical pump	The chemical pump could be set at a higher rate than required.
Leak	If the system is leaking water, it will use more chemicals. It's a good idea to install a water meter on the makeup water to the system
Carryover	If the system has carryover, the chemicals may be pulled into the piping and must be replaced.

Carryover is water from the boiler carried with the steam into the piping. It reduces the system efficiency, causes premature condensation of the steam, and uncomfortable areas.

You can check for carryover by looking at top of gauge glass. If water is coming from the top, the boiler has carryover. The water treatment specialist should be consulted if you suspect carryover.

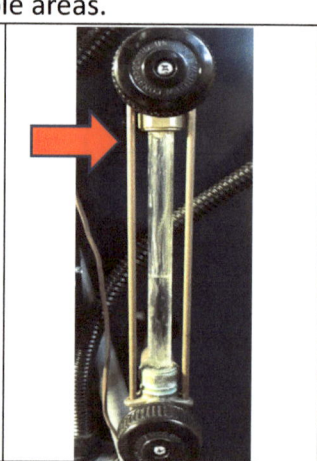

Percent of heat loss due to scale formation			
Scale thickness	Percentage of heat lost		
	Soft carbonate	Hard carbonate	Hard sulfite
1/50"	3.5	5.2	3
1/32"	7	8.3	6
1/25"	8	9.9	9
1/20"	10	11.2	11
1/16"	12.5	12.6	12.6

Boiler water treatment readings & possible causes

Findings	Effects
Excessive Hardness	Increases treatment costs & allows scale formations inside the boiler. Test softener operation.
Excessive Chlorides	Increases boiler water conductivity. Could indicate softener brine rinse not working.
Low PH	Can cause corrosion and iron deposits in the boiler and feedwater.
High PH	Could cause foaming and priming
Low Conductivity	Excessive blowdown which wastes fuel & chemicals.
Low Chlorides	
Low Silica in Boiler Water	
High Conductivity	Insufficient blowdown. Can cause scale deposits, foaming, and carryover.
High Chlorides	
High Silica	
Low Alkalinity	Can allow deposits and corrosion
High Alkalinity	Can cause foaming and carryover
Low Sulfite	Can allow corrosion
Low Phosphate	Can allow scale deposits
High Phosphate	Can result in phosphate deposits and could signify excessive chemicals in the system.
Low Molybdenum	Can result in scale deposits
High Molybdenum	Can indicate overfeeding of chemical treatment

Condensate water treatment readings & possible causes

High conductivity	Can indicate carryover
Low PH	Can cause corrosion and iron deposits in boiler and feedwater. May be caused by contamination of the boiler feedwater.
High PH	Can cause deposits
Iron	Indicates corrosion is occurring and leads to metal failure and iron deposits.
Excessive hardness	Leaks in condensate system

Water treatment chemical pump

Typical Water Quality Parameters in Steam Systems

Feedwater	Softness	Less than 1 ppm
Boiler Water	Hardness	Less than 1 ppm
	Ph	7-9 or 9.5-11*
	TDS	1,500-3000 ppm
		2,000-4000 µS
	Sulfite	30-60 ppm
	Hydroxyl Alkalinity	200 - 400 ppm
Condensate	pH	8.2-9.0
	TDS	Less than 20 ppm
*Verify with boiler manufacturer		

Condensate Return

Condensate tank

Boiler feed tank

Steam leaking from boiler feed tank vent, lots of money wasted.

Sludge in condensate tank with no strainer could destroy the feedwater pumps.

Always maintain an air cushion in tank. Tank should be between 1/3 and 2/3 full.

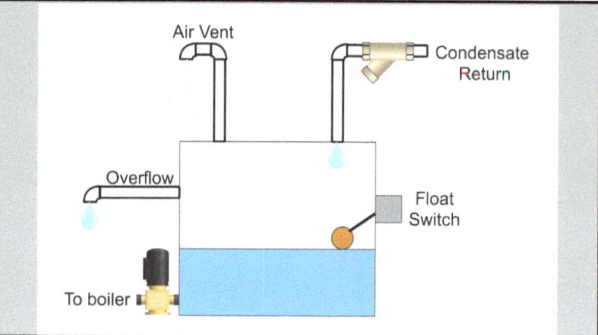

Condensate pipes are sized for 1/8 area of pipe for condensate

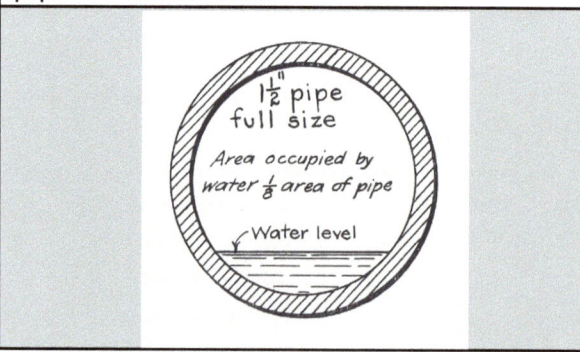

Condensate pump not working	
Power	Verify power is not interrupted. Check disconnect switch or breaker.
Incorrect voltage	Check for proper voltage.
Internal float is defective	Check internal float operation. It controls pump operation.
Defective pump	The pump could be frozen, or motor is defective.
Dirty condensate return strainer	If strainer is plugged or restricted, it may limit water flow into tank.
Pump lost prime	The pump may have lost its prime and is air bound.
Pump inlet plugged	Look for trash plugging the inlet to the pump.
Outlet pipe plugged / restricted	Verify pipe is not plugged or a valve closed

Feedwater pump is noisy	
Pump rotation	Check pump rotation. It could be running backward.
Pump pressure	The throttling valve may need to be closed more.
Humming	This could be from the pump motor wearing or locked up.
Excess condensate temperature	If the condensate temperature is too high, it will flash to steam inside pump
Bearing noise	Motor bearings may be worn, or motor is failing.
Relay chattering	Check for loose wire connections, low voltage, or burned contacts.

Condensate tank is leaking water through the overflow or vent.	
Defective float valve	The internal float valve operates the pump. If it's not working, the tank could fill with water.
No power	Verify there is power to the unit.
Defective condensate pump	Check condensate pump operation.

Steam is coming from the boiler feed tank vent	
Steam pressure too high	Excess steam pressure could cause the condensate to flash to steam and lost through the vent
Steam trap leaking	If a steam trap is leaking through, the steam is lost through the vent.
Master trap installed	A master trap installed close to the condensate or boiler feed tank could allow flash steam to enter the tank and out the vent.

Feedwater pump not operating	
Power	Verify power is available at the unit. Check disconnect switch or breaker if not.
Incorrect voltage	Check for proper voltage.
Low water cutoff not calling	If the low water cutoff is not working, it may not call for water.
Defective pump	The pump could be frozen, or motor is defective.
Excess amperage	The pump motor amperage may be greater than the amp capacity of the low water cutoff.
Pump lost prime	The pump may have lost its prime and is air bound.
Dirty condensate return strainer	If strainer is plugged or restricted, it may limit water flow into tank.

Boiler feed pump operates continuously	
Leaking steam traps	The condensate could be too hot and flashing to steam. The pump cannot pump steam.
Pump discharge pressure	The pump discharge pressure could be lower than the steam pressure.
Pump control	The pump control could be damaged.
Blocked discharge pipe	The pipe could be plugged, or the throttling valve is closed too far.
Check valve	The check valve could be installed backward or not working.
Pump too small	The pump is undersized.
Wrong rotation	The pump is running backward.
Blocked pump inlet	The pump inlet inside the tank could be blocked.

Feedwater pump capacity is reduced	
Pump capacity is reduced	Check for blocked pump inlet
Rotation	Verify pump motor rotation is correct.
Pump inlet blocked	Check for blockage in inlet to pump.
Discharge pipe	The discharge pipe from the boiler feed pump to the boiler could be plugged, restricted, or have a closed valve.
Pump discharge pressure	Verify pump discharge pressure is adjusted to design pressure.
Condensate temperature too warm	If steam traps are leaking, the pump will lose capacity as the water will flash to steam.

Boiler feed tank is flooded	
Water pressure to float valve is too high	A pressure reducing valve is used to limit the incoming water pressure to 25 psi
Defective float valve	Verify float works and shuts off makeup water valve.
Undersized tank	If the tank is undersized, it could flood
No overflow pipe	Boiler feed tanks require an overflow pipe to prevent the tank from flooding
System used to be a vacuum return	Vacuum return systems had condensate pipes smaller than gravity returns. You may need a larger tank
Check valve on pump discharge is leaking	If the check valve is leaking, water from the boiler could backflow into the tank. Try using spring loaded check valves.
Wet return plugged or restricted	If the wet return is restricted, it will slow the water returning and new fresh water will be made up into the system.

Condensate pumps are failing	
System dirt	Dirt and rust from the piping will end up in the boiler feed tank. Install a strainer on the condensate return pipes before tank. Flush tank at least once a year.
Leaking steam trap	This could elevate the condensate temperature and it flashes to steam inside pump volute.
Steam pressure too high	The higher the steam pressure, the hotter the condensate is. The water could flash to steam inside the pump volute.
Low water inside boiler feed tank	If the water level drops inside the tank, it could ruin the pump
Feedwater pipe is closed or restricted	Verify water is available to the boiler feed tank float valve.

Condensate pump is leaking	
Mechanical seal	If the pump is leaking water on a boiler feed or condensate pump, the mechanical seal is probably defective and should be replaced.

Honeywell contends a low high low burner is 15% more efficient than a modulating burner.

Gravity Returns

Gravity return systems use the vertical height in the "A" dimension pipe to overcome boiler steam pressure.

How do you get 1 1/2 psi into a boiler @ 2 psi?

1 3/4 psi

1 1/2 psi

2 Psi   2 Psi

1 1/2 psi

"A" Dimension 28" = 1 psi @ 2 psi steam pressure

1 3/4 psi

1 1/2 psi

2 Psi   2 Psi

28"   "A" Dimension

2 1/2 psi

How do you get water into boiler when using a zone valve? There is no left-over steam. In most instances, you may need a mechanical return such as a boiler feed tank.

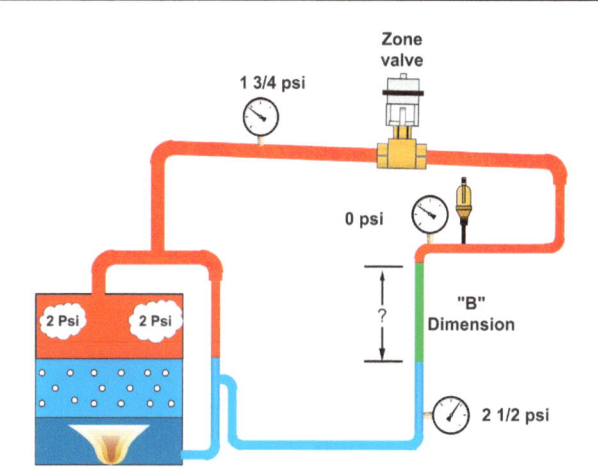

B dimension will have to be 70 inches high, and the pipes will be banging

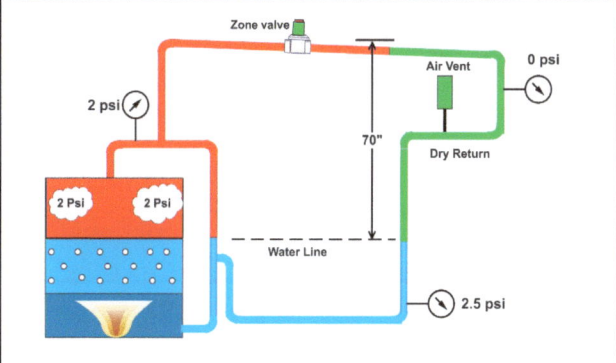

Understanding the "A" Dimension
As water rises in a vertical pipe, the weight at the bottom increases.

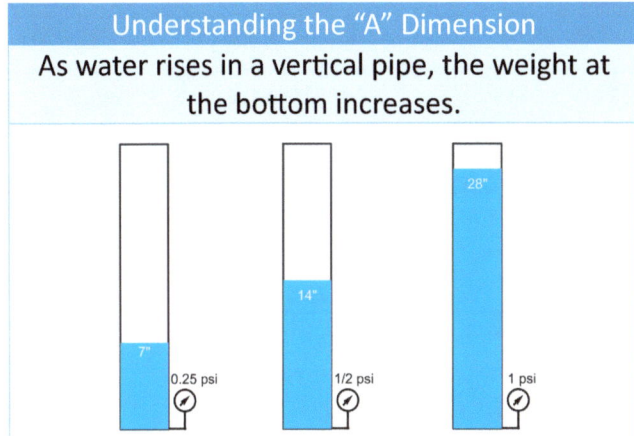

Problems outside the boiler room

Radiator air vent is panting	
Steam zone or radiator control valve	A closed valve could stop the air from leaving or entering. Consider a vacuum breaker.
System going into a vacuum	If the system goes into a vacuum, it will not allow condensate and steam to flow the proper way.
Radiator not pitched correctly	If the radiator is pitched incorrectly, water could pool inside the radiator.
Water is pooling	Look for areas where water may pool like a sagging pipe or eccentric pipe reducer.
Missing pipe insulation	Missing pipe insulation causes premature steam condensing.
Steam is condensing too quickly	Look for missing pipe insulation or steam pipes going through a cold area.

Radiator vent hissing or spitting	
Carryover	Carryover causes premature steam condensing.
Rad air vent not working	Vent could be partially plugged. Replace or clean air vent.
Main air vent not working or too small	If main air vent is defective or too small, the air in the system is vented through the radiator vents, making them noisy.

Radiator is dripping	
Valve packing is loose	Try tightening valve packing nut ¼ turn.
Radiator is leaking	Look for crack or hole in radiator.

Radiator air vent is noisy	
Main air vent not working	If main air vent isn't working, all the air will be vented through radiator vents.
Radiator vent is plugged	The air inlet /outlet could be plugged or restricted.

Radiator not heating	
One pipe air vent is defective	If the vent is defective, it won't allow air out and steam can't enter
Two pipe steam trap is defective	If the trap is failed closed, it won't allow air to leave and steam to enter.
No steam to radiator	Verify steam valves are open to the radiator. Verify radiator valve opens.
Condensate pipe is plugged or has a sag	Verify pipe is open and no sags where water can pool.
Radiator is enclosed	If the radiator is enclosed, the heat may not be able to escape.
Radiator doesn't heat all the way across	It may not need to if it's a milder day. The radiator only heats all the way on really cold days.

Radiator banging	
Valve partially open on a one pipe radiator.	Valve must be all the way open or closed on a 1 pipe radiator
Radiator is not pitched properly	Rad should be pitched to the valve on 1 pipe and toward the trap on 2 pipe systems
Water is trapped in radiator	Verify no water is pooling inside radiator

No heat in some areas	
Undersized or underfired boiler	If the boiler is too small, it will not send steam to far areas of building
Carryover	Carryover causes premature steam condensation
Steam pressure too low	If the steam pressure is too low, the steam may not reach the ends of the system
Air trapped	If air cannot be removed, steam can't enter.
Is radiator covered?	Covering the radiator reduces the heat output. Change or remove cover.
Incorrect location of thermostat	Is the thermostat getting a false reading from sun shining on it or steam pipes behind wall?
No air space in boiler feed tank	If the condensate or boiler feed tank is flooded, it will not allow air to enter or leave the system.
Sagging pipe	Water pooling in a horizontal pipe could cause premature steam condensation.
Condensate return pipe is plugged or restricted	If the condensate return is restricted or plugged, the system can't breathe.
Missing pipe insulation	Missing pipe insulation causes premature steam condensing.
Failed steam trap	Sometime steam traps fail closed and stop steam from entering radiator.
System can't breathe	Walk around system and verify air can enter or leave the system.
Air venting too quickly in radiator	If the air vents too quickly, the steam may change direction and go toward the fast air vent.
Leaking steam trap in another area	A steam trap leaking in another area could pressurize the condensate

	pipe resulting in air trapped in radiator.
Someone installed one pipe air vent on a 2-pipe system radiator	A one pipe air vent on a two-pipe system causes steam system issues. It also masks whether the steam trap on the radiator is defective.
Strainer plugged	If the strainer is plugged, steam or condensate may not get through.
Different heat emitters	If one part is copper baseboard radiation and the other is cast iron radiators, there will be issues. Copper heats & cools quickly while cast radiators heat & cool slowly. Its best to have them on separate zones.

Odor coming from radiator air vent	
Carryover	Carryover could bring the boiler water with the steam.
Boiler needs cleaned	If boiler wasn't properly cleaned, you could be smelling the oil from pipe threading.
Water treatment	If someone added water treatment to the boiler, the odor could come out the vent.
Water is dirty	Try skimming or cleaning the boiler.
Oil is in water	Pipe threading oil will smell foul when air is vented.

Radiator gurgling	
Incorrect pitch on radiator	Water pooling inside the radiator causes premature steam condensing. Check for proper radiator pitch.
Carryover	Carryover causes premature steam condensing.

Heat takes too long to get to radiator

Sagging pipe	A sagging pipe could cause a water seal, trapping air and stopping steam flow.
Radiator vent partially closed or restricted	If the valve is plugged or partially closed, it would be slow to heat.
Air in system	If air is in system, steam can't enter. Look for broken or missing air vents.
Main air vent too small	Steam systems are designed for the steam to reach the ends of the run in 60 seconds. Try adding an additional air vent.
Radiator air vent not working	Verify the air vent on a 1 pipe system is working. Clean outlet or replace if not.

System stops heating when steam is formed in piping

Incorrect air vent	If the steam pressure is too high, it may not open unless the steam drops below the drop away pressure of the air vent.

Main air vent leaking steam

Defective air vent	The air vent may need replaced.
Improper installation	If the air vent is installed too close to the end of the steam main, the water hammer could destroy it.
Water hammer	Check for water pooling in pipe or missing pipe insulation.

> Typical pressure settings for a 1 pipe system: Start @ ½ psi and stop @ 2 psi.

Steam doesn't heat furthest radiators

Undersized or underfired boiler	If the boiler is underfired, there may not be enough steam to reach the furthest radiators
Air bound	If the pipe or radiator is filled with air, steam cannot enter.
Air vent too small	An undersized air vent will slow the steam to the area.
Missing pipe insulation	If the pipe insulation is gone, the steam will collapse prematurely
Thermostat in wrong location	If the thermostat is sensing heat, it may shut off before steam reaches the cold area.

Main air vent is failing

Air vent not piped correctly	Verify air vent is piped correctly. It should be 15 inches from last elbow and 6-10 inches high
If air vent is too close to the 90 looking down, the condensate will hit the elbow and destroy the vent.	

Unbalanced heat in building

Air vent is too small	If the air vent is too small or not working, the air can't get out so steam can't get in. Verify air vents work.
Water seal in piping	Check for sagging pipes where a water seal could form
Valve closed	Verify valves are open to the system

Intermittent heat in some areas	
Steam trap too far from radiator or coil	This could cause steam to be caught inside piping. Move steam trap closer to radiator or heating unit.

AHU Steam coil is freezing	
No vacuum breaker	Install vacuum breaker between control valve and steam trap
Trap is too close to coil	Water gets trapped in coil. Lower steam trap to 28" below coil.
Trap sized incorrectly	Verify trap is sized for .025" differential pressure

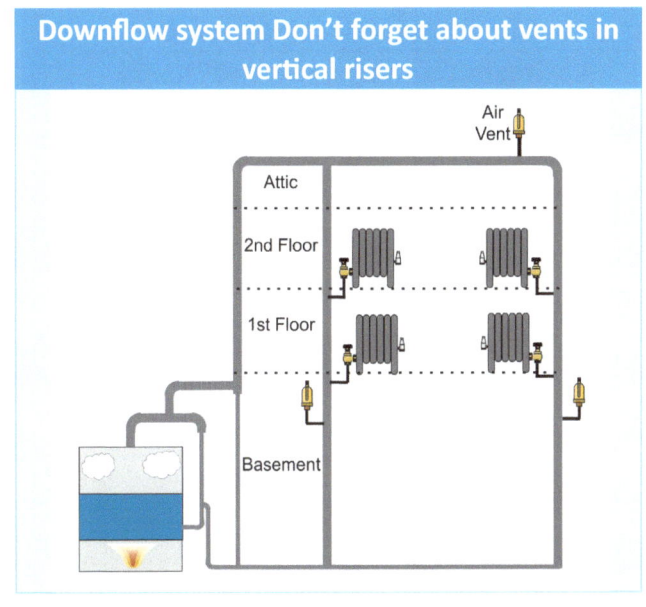

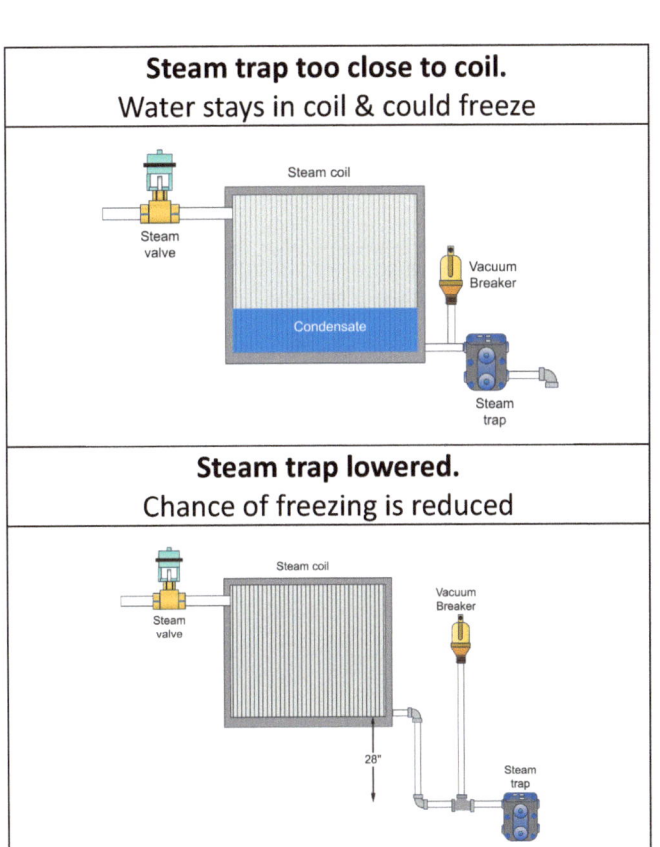

Steam trap too close to coil.
Water stays in coil & could freeze

Steam trap lowered.
Chance of freezing is reduced

Steam traps were sometimes used to vent air from the steam pipe to the dry return.

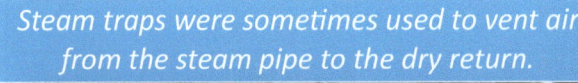

Scale forms on the hottest surfaces.

Steam Traps

Steam Trap Sizing Rules of Thumb Source: BarnesandJones.com	
Steam Mains	Safety factor when sizing Low Pressure 2 X Actual Load, High Pressure 3 X Actual Load
Distance between traps	500 feet if supervised startup
	200 feet if automatic start-up
Trap Locations	At all low points
	At each change in elevation
	Before all control valves
Size of collection pocket for drain trap	Full size pipe diameter tees up to 12"
	If pipe is over 12" use one pipe size smaller
	Length of drip 1 ½ times pipe diameter but not less than 8"
Velocity in steam main	Low Noise 4,000 to 6,000 FPM
	Industrial plant 8,000 to 12,000 FPM

Startup Condensate Loads from 70 Degrees F Schedule 40 Pipe Pounds of Condensate per Linear Foot			
Pipe Size Inches	Weight per foot of pipe	Steam Pressure Psig	
		2	15
1	1.69	.030	.037
1 ¼"	2.27	.040	.050
1 ½"	2.72	.048	.059
2"	3.65	.065	.080
2 ½"	5.79	.104	.126
4"	10.79	.190	.234
6"	18.97	.335	.413
8"	28.55	.504	.620
10"	40.48	.714	.880
Based on 90% efficient insulation			

Compare steam trap applications Source Armstrong			
	Thermodynamic Traps	F&T Traps	Inverted Bucket
Modulation	Fair	Good	Good
Backpressure	Poor	Good	Good
Dirt	Poor	Poor	Good
Wear	Poor	Good	Good
Water Hammer	Good	Poor	Good
Freezing	Good	Poor	Good

When trapping the steam main, the drip leg should be same pipe size as the steam header diameter up to 4 inch.

Correct	Incorrect

A master trap is a steam trap installed before the condensate or boiler feet tank. It is not a good idea. It causes excess flash steam, increases operating costs, and reduces system comfort.

A burner should operate for 15 minutes to allow the flame to stabilize before doing air to fuel adjustments.

Float & Thermostatic Steam Trap - Troubleshooting

Steam trap discharge is cool	
Verify steam valve is open & steam is available	Open valve or start boiler
Upstream strainer is plugged	Clean or replace screen
Blockage or water seal in condensate pipe	Remove blockage or fix sagging pipe
Verify trap is not plugged	Disassemble and clean inside trap

Steam trap is blowing steam	
Check for dirt inside trap	Clean and reassemble
Check thermostatic element	Replace or rebuild trap
Worn components inside trap	Replace or rebuild trap
Verify a different trap isn't blowing through	Find and replace or rebuild the leaking trap

Uneven or slow heating	
Thermostatic element defective	Replace or rebuild trap
Blockage or water seal in condensate pipe	Remove blockage or fix sagging pipe
Verify steam trap is sized properly	Replace with properly sized trap
Trap too small	Replace with proper size trap.

Thermostatic Steam Trap Troubleshooting

Steam trap discharge is cool	
Check bellows inside trap	Replace bellows or trap
Verify steam valve is open & steam is available	Open valve or start boiler

Steam trap discharge is hot or blowing steam	
Check bellows inside trap	Replace bellows or trap
Trap seat is dirty	Clean dirt from trap

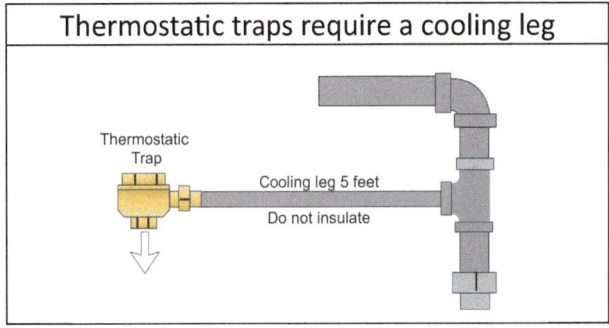

Thermostatic traps require a cooling leg

Calculate condensate loads	
Heat water with steam =	Pds/Condensate/Hr = (GPM ÷2) x Δt
Heat air with steam coils =	Pds/Condensate/Hr = (CFM ÷900) x Δt
Steam Radiation =	Pds/Condensate/Hr = EDR Sq. Ft ÷ 4

Don't allow water to pool against a zone valve unless you like banging pipes

Typical Combustion Readings for Category I boilers

Atmospheric Gas Burner	
Oxygen	7% - 9%
Stack Temperature	325 to 500^{0}F
Draft in WC Inches	-.0 "WC to -.04" WC
Carbon Monoxide PPM	<100 PPM Air Free

Gas Power Burner	
Oxygen	3% - 6%
Stack Temperature	275 to 500^{0}F
Draft in WC Inches	-.02" WC to -.04" WC
Carbon Monoxide PPM	<100 PPM Air Free

Oil Power Burner	
Oxygen	4% - 7%
Stack Temperature	325 to 600^{0}F
Draft in WC Inches	-.04" WC to -.06" WC
Carbon Monoxide PPM	<100 PPM Air Free with No Smoke

Carbon Monoxide (CO) Exposure Effects	
CO PPM	**Effects**
200	Slight Headache, Tiredness, Dizziness, nausea after 2-3 hours.
400	Frontal Headaches 1-2 Hrs., Life Threatening After 3 Hours.
800	Dizziness, Nausea & Convulsion within 45 minutes. Unconsciousness within 2 hours. Death Within 2-3 Hours.
1,600	Headache, Dizziness & Nausea within 20 minutes. Death within 1 Hour
3,200	Headache, Dizziness & Nausea within 5-10 minutes. Death within 30 minutes
6,400	Headache, Dizziness & Nausea within 1-2 minutes, Death within 10-15 minutes
12,800	Death Within 1-3 Minutes

Carbon monoxide dangers

Negative condition in boiler room If you look at the pictures below, the top picture is a normal boiler room, and the combustion air opening allows the boiler to operate safely. The lower picture shows how a boiler room can go negative allowing the dangerous flue gases to be pulled from the

chimney and enter the boiler room and the building. It only takes -0.02" WC negative pressure to pull the flames from a burner. Exhaust fans in a boiler room could be dangerous.

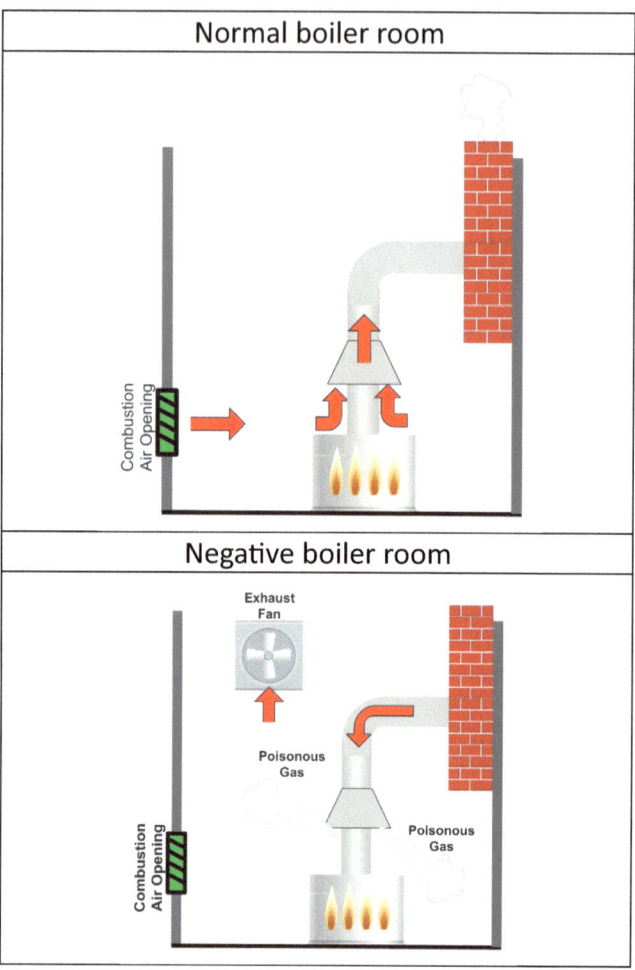

Water expands at 3% of its volume when heated from 32^{0}f to 180^{0}F

Understanding the Boiler Sequence of Operation

Atmospheric burner

- The electrical power is switched on and there is a call for steam.
- Presence of water is verified.
- The pilot flame is ignited and verified. Most atmospheric burners use a flame rod placed inside the pilot flame to verify the pilot flame lights.
- Once the pilot is verified by the flame safeguard, a signal is sent to open the gas valve(s).
- The burner will stay lit until either the call for steam ends, the steam pressure rises high enough to meet the setting of the steam pressure control, or one of the safety controls opens.

Power burner On Off

- The electrical power is switched on and there is a call for steam.
- Presence of water is verified.
- The burner blower motor starts & verified, typically using a differential pressure switch.
- Once the blower is verified, the burner runs 30-90 seconds to purge any leftover fuel inside the boiler or chimney. This is called Pre-Purge.
- At the end of the pre-purge, the ignition transformer is energized, and the ignition electrodes starts to spark. The ignition transformer is rated for 6,000 to 12,000 volts. At the same time, the pilot solenoid valve opens, and the spark ignites the pilot flame. The pilot flame is verified using a flame rod, infrared, or ultraviolet sensor.
- After the pilot operation is verified, the main gas valve(s) open.
- The burner operates until either the call for steam ends or the steam pressure raises to the pressure control setpoint.

Typical Sequence for Boilers with Variable Stage Power Burners

Low High Off Power Burners After the pre-purge, the burner will travel to low fire & ignite & verify the pilot. When the pilot is verified, the flame safeguard will open the electric shutoff valves. Once the flame is established, the burner drives to high fire and will stay there until the call for heat ends or the boiler pressure meets the pressure control setpoint.

Typical Sequence for Low High Low Power Burners

After the pre-purge, the burner will drive to low fire, ignite & verify the pilot flame. When the pilot is verified, the flame safeguard will open the electric shutoff valves. After the flame is established, the burner drives to high fire to meet the setpoint of the firing rate control. Once the steam pressure reaches the setpoint of the firing rate control, it will then drop to low fire. The burner will travel between low and high fire as directed by the firing rate control until the call for heat has ended or the boiler pressure reaches the pressure control setpoint.

Typical Sequence for Modulating Power Burners

After the pre purge, the burner drops to the low fire position & ignites & verifies the pilot flame. When the pilot is verified, the flame safeguard will open the electric shutoff valves. After the flame is established, the burner drives to high fire to meet the setpoint of the modulating control. When the steam pressure approaches the modulating control setpoint, the control will position the burner between low and high fire to maintain the steam pressure. It will continue adjusting the firing rate until the call for heat ends or the steam pressure meets the operating control setpoint.

The higher the steam pressure, the slower the steam travels.

Brewery Calculations

Steam systems in breweries are different than their space heating counterparts. For one thing, the steam pressure requirements are higher than for space heating. Many of the brewery equipment manufacturers want 15 psi steam pressure. It's almost impossible to deliver that pressure on a low-pressure steam boiler. If you try, you risk tripping the manual reset limit control which has a top setting of 15 psi. If the limit control trips, the boiler will not start again until the pressure drops below the setpoint, and the manual reset button is pushed.

That pressure, 15 psi, is also the setpoint for the pop safety valve. If the pop safety opens, the customer loses all the steam, and the process comes to an abrupt stop. I like operating the boiler between 11 and 13 psi.

Another common issue is the boiler should be sized for the connected load. If the brewer is using a twenty-barrel system and only brewing ten, the boiler should be sized for the twenty-barrel load.

Kettle or Mash Tun does not heat	
Steam pressure	Verify boiler steam pressure is at the pressure you require.
Control valve	Check to see if the control valve opens on the kettle or mash tun.
Manual valve	Visually inspect the piping to see if any manual valves are closed.
Trapped air	Look for a sagging pipe after the steam trap to be sure there is not a water seal which would not allow air to escape.
Steam trap	Test the steam trap on the outlet to see if it works.

Steam pipe sizing capacity for 12 Psig steam pressure @ 40 FPS or Feet per Second		
Input MBH	Output MBH	Minimum Pipe Size In
200,000	160,000	2
300,000	240,000	2 1/2
400,000	320,000	2 1/2
500,000	400,000	4
600,000	480,000	4
700,000	560,000	4
800,000	640,000	4
900,000	720,000	4
1,000,000	800,000	4
1,100,000	880,000	4
1,200,000	960,000	6
1,300,000	1,040,000	6
1,400,000	1,120,000	6
1,500,000	1,200,000	6
1,600,000	1,280,000	6
1,700,000	1,360,000	6
1,800,000	1,440,000	6
1,900,000	1,520,000	6
2,000,000	1,600,000	6

Btu Requirements for Craft Brewer Based on above formulas by Blain Clouston			
Barrels	Brewkettle	Hot Liquor	Mash Vessel
1	32,000	19,500	29,000
5	160,000	97,500	145,000
7 ½	240,000	146,250	217,500
10	320,000	195,000	290,000
12 ½	400,000	243,750	362,500
15	480,000	292,500	435,000
17 ½	560,000	341,250	507,500
20	640,000	390,000	580,000
22 ½	720,000	438,750	652,500
25	800,000	487,500	725,000
27 ½	880,000	536,250	797,500
30	960,000	585,000	870,000

It takes less than a second to get third degree burns from steam.

Brewery Formulas
Brewkettle heating- 1.5-degree F/minute
Brewkettle, evaporation- 5- 6%/hr
Mash Vessel Heating- 1.5 degrees F/minute
Hot Liquor tank- ability to heat water 1.5 times the brewhouse size in 4 hours. Example: a 10-barrel system would typically use up to 15 barrels in hot liquor.

Brewery Btu's Required	
Item	Btu/Hr/Bbl.
Brewkettle Heating and Boiling	32,000 Btu/Hr/bbl.
Hot Liquor Heating from 60-175F in 4 hours =	19,500 Btu/Hr/bbl.
Mash Vessel Heating =	29,000 Btu/Hr/bbl.

Formula to calculate Btu's required to heat liquids in steam jacketed kettles
$$Lbs\ Condensate/hr = \frac{G \times sg \times Cp \times \Delta T \times 8.3}{L \times t}$$
G = Gallons of liquid to be heated sg = Specific gravity of the liquid Cp = Specific heat of liquid ΔT = temperature rise of the liquid L = Latent heat of the steam (Btu/lb.) T = Time in hours
Specific Gravity Water = 1.00 Specific Gravity of Wort = 1.050 @ 60^0F Specific Heat of Water = 4.219 @ 212^0F Specific Heat of Wort = 4.0 – 4.1 Specific Heat of Mash (15^0P) = 3.73 Specific Heat of Mash (20^0P) = 3.60 Specific Heat of Mash (25^0P) = 3.46 Specific Heat of Air = 1.005

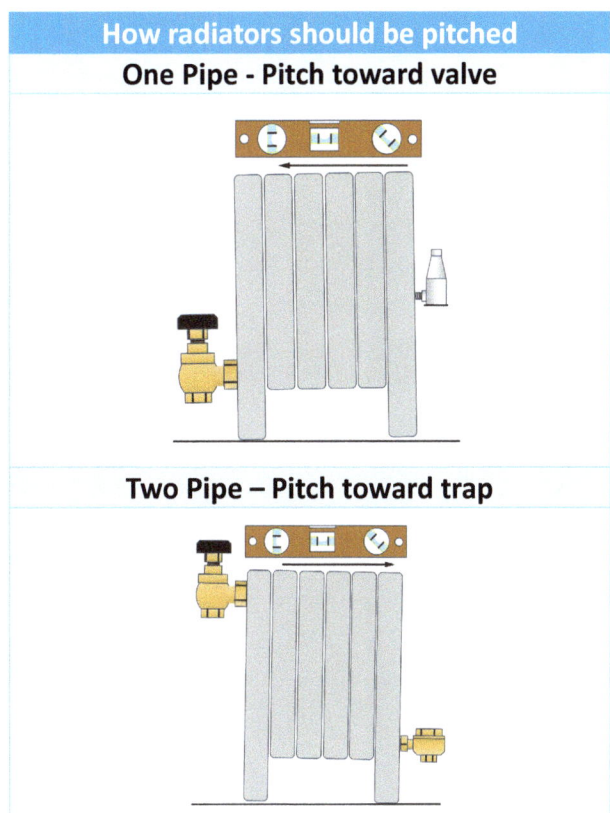

How radiators should be pitched

One Pipe - Pitch toward valve

Two Pipe – Pitch toward trap

Two gases cannot occupy the same space.

Evaporation test of low water cutoff

More insurance underwriters and government agencies are requiring an *Evaporation Test* be performed on steam boiler low water cutoffs. This test simulates actual operating conditions. While it may or may not be required in your area, it is a recommended test.

When testing the low water cutoffs by opening the bottom blowdown valves, the force of the boiler steam pushing against the top of the float may force the float down, shutting off the boiler. During normal operation, the pressures above and below the float are closer.

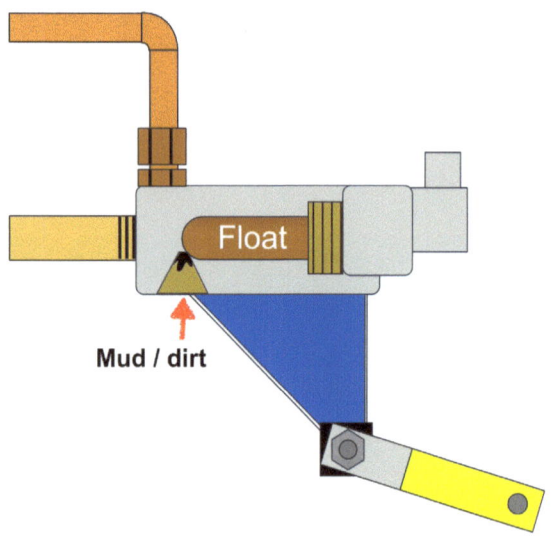

Mud / dirt

Dirt or mud inside the low water cutoff could impede the float from dropping under normal operations but does so when doing the blowdown. The evaporation test will show that. Check with the boiler manufacturer on how they suggest the test be done. This is how I do it.

Performing the evaporation test on a commercial boiler with a feedwater pump is easier than doing one on a residential boiler but still takes time. I shut off all the boilers, so no steam from the other boiler fills the boiler you want to test. Next, I shut off the feedwater pumps feeding the boiler and close the manual valve. After I start the boiler, I want to test, I wait and watch the gauge

glass. Make sure not to leave the boiler unattended during this test.

As the boiler fires, steam will leave the boiler, and the water level will drop. If it falls below the glass in the gauge glass, there is something wrong, and the boiler should be shut off and the low water cutoff disassembled. Check the float, clean any debris found in the safety control, and repeat the test when done. If it does it again, replace the control. Don't forget to open the valve and start the feedwater pump when the test is complete. Repeat this test on the other boilers.

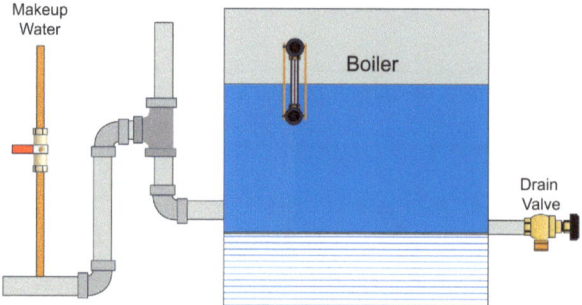

On residential steam systems, I shut off the makeup water pipe. Theoretically, the system should never require makeup water. This assumes there are no sagging pipes where water could pool, and the wet return is open and clear. I know, this is a big assumption. To help the test along, I put a hose on the boiler drain and direct the other end into the drain. Once the boiler is steaming, I will partially open the drain valve and let the water trickle out. Then I watch the gauge glass. If the water level drops below the gauge glass, I shut off the boiler. I open the low water cutoff, clean the bowl, and check the float. If it does it again when retesting, replace the control. Open the makeup water line and start the boiler when satisfied. Bring hose caps with you, as the drain valve will probably leak. The evaporation test should be done every year.

Suggestions to avoid problems on your next steam boiler replacement.

Read the instruction manual. There are lots of gotchas sprinkled through it which could affect the boiler and system operation.

Near Boiler Piping

Correct near boiler steam piping is crucial. Follow the manufacturer's suggestions.

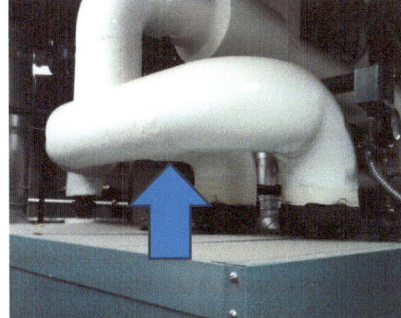

Swing joint will absorb the expansion difference between the pipes & boiler.

The vertical pipe or "A" dimension should be at least 28" above the boiler water line to push the return into the boiler when firing @ 2 psig.

If ceiling height is low, consider using a drop header.

Insulate steam pipes. It will prevent the steam from condensing prematurely.

Drop header

The Hartford Loop connection should be a close nipple and connected 2-4 inches below the boiler water line.

System Piping

Flush or replace the wet return piping. It will plug over time and has an estimated life of ten to fifteen years.

Use schedule 80 black iron pipe for condensate piping, it's 50% thicker than schedule 40.

Insulating condensate pipes reduces carbonic acid.

Verify steam piping is pitched properly.

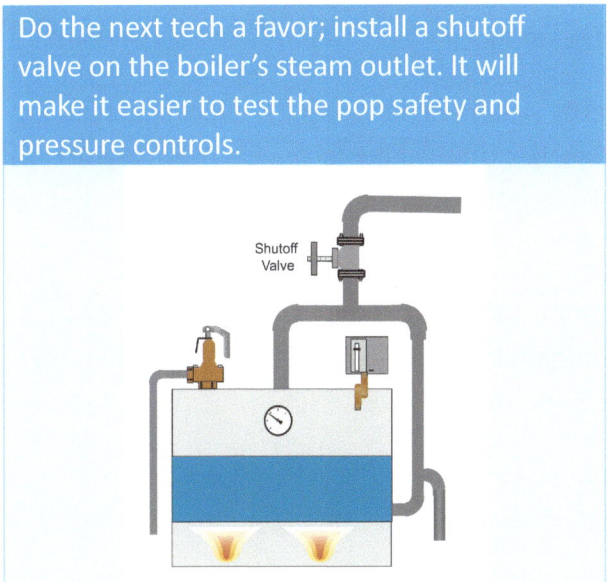

Do the next tech a favor; install a shutoff valve on the boiler's steam outlet. It will make it easier to test the pop safety and pressure controls.

If water pools inside the pipe, the steam prematurely condenses and lowers efficiency.

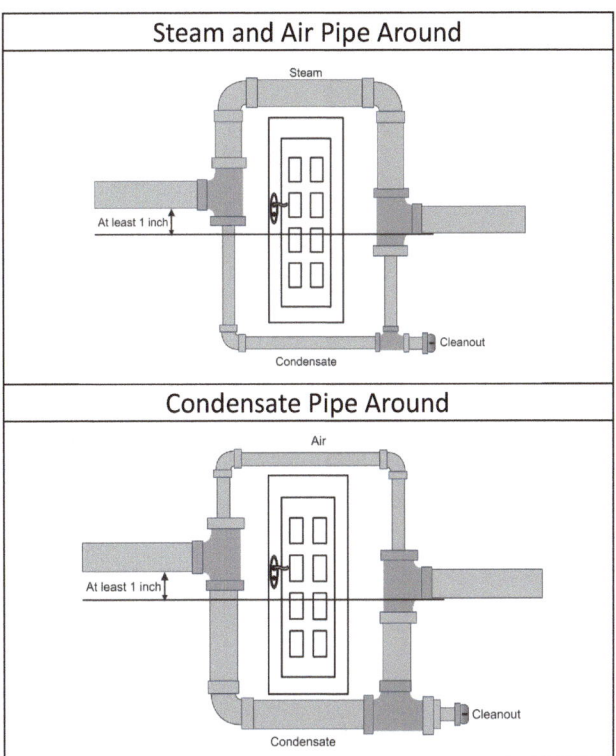

Some manufacturers require threaded joints in the steam piping to absorb the expansion and contraction of the piping. Check the manual.

Install a vacuum breaker if using automatic steam valves.

Trap the pipe when using a zone valve so water does not lay against the valve seat. It will bang when it opens.

Verify system main air vent is in proper location, 15 inches from elbow facing down and 6-10" high. Replace the vent when replacing the boiler.

Boiler

Commercial steam boilers require two low water cutoffs & two pressure controls: one automatic and one manual reset.

Steam space heating systems should be set for 2 psig. Brewery boilers should be set @ 11 psig.

Size the replacement boiler for the connected load and not by the building's heat loss.

Clean boiler after installation.

Pop safety valve discharge pipe should not have threads where someone can install a pipe cap.

Verify the pop safety valve discharge piping is not reduced or plugged.

When connecting to the building's EMS system, have the building control system just enable or disable the heat. Use the boiler's controls to adjust firing rate.

Old atmospheric boilers used 3 ½" WC gas pressure. New boilers may require 7-14" WC.

Remember to vent the gas train components outside.

The new boiler will likely use higher amps than the atmospheric boiler.

Condensate Return

The mechanical return system for a boiler should be a boiler feed system and not a condensate tank. All makeup water is piped into boiler feed

42

unit.

Use a pressure reducing valve set for about 25 psi on the makeup water to the boiler feed tank.

A spring-loaded check valve on the boiler feed pump discharge works better than a swing check.

Use a strainer on the condensate return pipe to the condensate or boiler feed tank. It will extend the feedwater pump life.

The vent pipe on a boiler feed tank should never be reduced or plugged.

Verify condensate piping to tank does not have a water seal to impede air flow.

A spring-loaded check valve on the pump outlet is a better choice than a swing check to avoid a vacuum in the boiler from pulling water from tank into the boiler.

Overflow & vent pipe should be same diameter as the tapping on the tank. They should not be plugged or reduced.

Check the instruction manual for the pump to see if bleed valve connected to ¼" tubing should be open or closed. This is for the mechanical seal.

Water Treatment

All steam boilers should have water treatment.

A water softener on the makeup water reduces scale formation.

Install a water meter on the makeup line to

monitor system water loss. Residential steam systems should lose no more than a gallon a year. Commercial steam systems should return 95-97%.

Install a skim pipe on the boiler just below the water line. Most TDS inside a boiler are within 6-8" of the water line.

Repair all steams system leaks.

Venting the Boiler

The horizontal flue pipe can be no longer than 75% of the vertical chimney when using single wall pipe. It can be the same length as the chimney height when using "B" vent.

Verify flue is pitched up toward the chimney.

A draft control, like a barometric damper, is used when the chimney is over 30 feet tall.

Remove red stops on barometric damper when only firing with natural gas. The red stops are only used when firing with fuel oil.

Verify chimney dimensions are large enough.

Boiler Room

An emergency door switch just inside the boiler room door is a great idea and the code if your municipality follows ASME CSD1 code.

Instruct customer about risks of storing dangerous chemicals in boiler room.

Water hotter than 140^0F should not be piped into the drain. Instead, use a blowdown tank.

Blowdown tank

Verify adequate combustion air to the boiler room or basement.

Check combustion of the new unit and document what you found.

A master trap before the condensate or boiler feed unit could increase fuel costs, decrease comfort, and destroy the condensate pump.

Thermostatic steam traps need a cooling leg five feet before the trap.

When installing two or more boilers, install a high-level spill trap on each boiler to avoid flooding the idle boiler.

Outside the Boiler Room

Replace or rebuild the system steam traps. The life expectancy of a steam trap is 6 to 10 years.

Verify radiators are not covered. 60% of the heat from a radiator is convection. The cover should allow the free flow of air.

> The higher the pressure, the smaller the steam volume.

Why do boiler maintenance? There are two reasons for performing maintenance on a steam boiler: safety and longevity. The odds of a boiler accident drop when maintenance is done. The second reason for doing maintenance is longevity; boilers which are maintained last much longer than ones which are not. According to Hartford Steam Boiler, the three leading causes of steam boiler failures are due to Low Water, Scale, or Corrosion. If you do maintenance, these issues significantly diminish. Check with the manufacturer of your boiler for their requirements. The following is a generic one I use for most low-pressure steam boilers.

Daily Tasks
Look around the boiler room for unsafe conditions. Maintain 30" clearance around the boilers.
Look for signs of flame rollout from boiler or water heater.
Look for steam/water leaks around the boiler and piping. The leaks should be repaired ASAP.
Note boiler pressure when running.
Look for rust under the boiler or water heater draft diverter. This could signify back drafting of the flue gases.
Note gas pressure at the boiler. The gas pressure should be at the required pressure the manufacturer requires.
Sniff for natural gas leak. Repair gas leaks asap.
Watch and listen to the blower or induced draft motor. Failing motor bearings may be louder when motor starts or stops.
Watch the firing rate / modulating motor. Verify it works smoothly without binding.
Check the water level in the boiler gauge glass. If the water level is not visible, the boiler should NOT be fired. Investigate the cause.
Blow down boiler and steam column. Confirm low water controls shut off the burner and the pump control starts the boiler feed pump. This is done at low fire under light conditions. If the low water controls do not shut off the burner, the boiler should not operate until the problem is repaired.

Check chemical feed and softener system operation.
Visually inspect burner flame & verify the flame is impinging on the metal surfaces.
Be sure the combustion chamber is not covered in soot.
Check the boiler flue draft. It should be slightly negative, in most instances.
Verify pop safety valve isn't leaking.
Check boiler pressure gauge. The pressure should be around 2 psig for space heat & 11 for a brewery.
Check stack thermometer when boiler is running. If its higher, this could indicate a dirty fire side or scale on water side.
Verify pipe insulation is intact & no evidence of a leak. Wet insulation should be changed.
Inspect stack. Look inside when boiler is NOT firing to see if soot is forming in the flue.
Verify flue pipe is intact, no deterioration or blockage, or rust holes. Verify it's connected properly and pitched up to the chimney.
Be sure the combustion air openings are free and clear.

Weekly Tasks
Check water meter on the makeup water pipe. Note weekly usage and investigate if water usage increases.
Check supply levels for water treatment and water softener. Be sure you have enough supplies to last the week.
Check boiler/burner light bulbs. Replace burnt out bulbs.
Regenerate water softener.
Check fuel train for leaks. Look at window in gas valve actuator for hydraulic fluid leak. Replace the actuator if leaking.

A water drip per second equals 8 gallons per day or almost 3,000 gallons per year lost.

Monthly Tasks

Check gauge glass for damage, wear, or etching. Replace if you notice those things.

Check the linkages on gas train components. Look at linkages and ball joints for wear or slippage.

Check gas train venting. Verify gas train components are properly vented and the vents are connected

Twice Yearly Tasks

Power burner. If the boiler has a power burner, it should be checked twice per year and the fuel to air ratio should be tested and adjusted.

Perform slow evaporation test. If the burner does not shut off, the boiler should be shut off and not operated until the problem is repaired.

Annual Tasks

Open water side of boiler and look for corrosion, scale, electrolysis, or rust. Investigate further if any is found. Clean waterside of boiler. Remove scale and flush boiler with hose.

Open fireside of boiler and look for water leaks into fireside. Check refractory for cracks and missing pieces. Repair or replace missing or damaged refractory. NOTE: If refractory gets wet, it should be thoroughly dried before exposing it to flame. Entrained moisture can cause it to break apart when heat is applied.

Open low water cutoffs and inspect probes and floats. Replace any defective or worn components.

Inspect sensing pipes for steam pressure controls. Verify the piping and pigtails are free of obstructions and clear.

Open wye strainers. Clean any sludge buildup and clean screens

Replace gauge glass. This would include gauge glass, washers, and rings.

Flush boiler feed and condensate tanks

Verify boiler feedwater pipe is free and clear.

Replace water side and fire side gaskets

Check burner blower motor. Verify it is clean and operates without excess vibration.

Fill boiler with water and check for leaks.

Check and adjust air to fuel ratio. Use a calibrated combustion analyzer to adjust the air to fuel ratio. This should be a task done by someone trained on boilers.

Check steam pressure controls & pigtails.

Open piping connecting safety controls to boiler. Verify the sensing pipes are clear.

Leak test the gas train.

Check flame safeguard according to the manufacturer.

Test the boiler pop safety valve. Typically, the boiler should be at 75% of the rated pressure setting. On a low-pressure steam boiler, the pop safety valve should be tested once the boiler steam pressure reaches 11 psi or higher.

Check all wiring terminations. Verify all the screws are snug on the wiring terminals. Be sure the power is off when doing this.

Check boiler base on atmospheric boiler. Look for missing or cracked refractory or damage to the base. Do not operate until it is repaired.

Check burners. Clean if dirty.

Condensate induced water hammer can generate pressures of 1,500 psi.

Getting your boiler ready for the boiler inspector

If your boiler requires an internal inspection.

Be sure the area around the boiler is free and clear and there is a minimum of 30-36" clearance around the boiler & burner.

Close main gas valves on the gas or fuel oil train

Drain boiler.

Lock off boiler with lock out tag out.

Open water side of boiler. Remove scale and flush boiler with water hose.

Open fireside of boiler and look for water leaks. Check refractory for cracks and missing pieces.

Open low water cutoffs and inspect probes and floats. Replace any defective or worn components. When leaving the low water cutoff disassembled for the boiler inspector, do not allow the control to hang by the wire. This could cause the wiring to fail or break. Use a rope or chain to hold the control in place so there is no stress on the wiring connections.

Disassemble and inspect sensing pipes for steam pressure controls. Verify the piping and pigtails are free of obstructions and clear.

Open and remove manholes and handholes from boiler if applicable. Look inside and clean scale and dirt.

Remove pop safety valve and verify pipe feeding the safety valve to be free and clear.

Flush the boiler mud legs with a water hose.

Verify boiler feedwater pipe is free and clear.

After the inspection

Replace water side and fire side access panels or doors. Use new gaskets.

Reinsert handholes and manholes. Use new gaskets

Reassemble low water cutoff sensing pipes and controls. Use new gaskets. Be sure the surfaces are cleaned.

Fill water in boiler and check for leaks. Look at fittings and see if tubes are leaking. All leaks should be repaired.

Complete startup of boiler which includes testing the safety and limit controls. Check and adjust combustion efficiency.

Flame Safeguard Signals

Honeywell	
Model	Average Flame Signal
RA890	
R4795	
R7795	2-6 µA DC
R4140	
R4150	
BC7000	
RM7890	
RM7895	1.25-5 VDC
RM7840	
RM7800	
S8600	1-5 µA DC
Fireye	
Model	Average Flame Signal
UVM	4.0-5.5 VDC
TFM	14-17 VDC
D-10/20/30	16-25 VDC
E-100/ 110	20-80 VDC
E-100/E110 with EPD Programmer	4-10 VDC
M Series II	4-6 VDC
Micro M Series	4-10 VDC
Micro M Series w Display	20-80 VDC
Typical Flame Signals	
Cad Cell	Less that 1,600 ohms
Flame Detector	Average Flame Signal
Flame Rod	2-5 µA DC
Flame Rod Self-Checking	1 ¼ - 2 ½ µA DC
Infrared	2 ¼-5 µA DC
Photocell	2-5 µA DC
Photocell Self-Checking	1 ¼ - 2 ½ µA DC
Thermocouple Q309	30mV
Thermopile Q313	750 mV
Ultraviolet	2-6 µA DC

Please verify with manufacturer

94% dry steam @ 200 psig has the same latent heat as 98% dry steam at 0 psig.

Heating Formulas

Steam

Properties of Saturated Steam				
Psia	Psig	°F	°C	Volume. Cu. Ft.
14.7	0	212	100	26.8
15.7	1	216	102	25.2
16.7	2	219	104	23.5
17.7	3	222	106	22.3
18.7	4	225	107	21.4
19.7	5	227	108	20.1
20.7	6	230	110	19.4
21.7	7	232	111	18.7
22.7	8	235	112	18.4
23.7	9	237	114	17.1
24.7	10	240	115	16.5
25.7	11	242	117	16
26.7	12	244	118	15.3
27.7	13	246	119	15
28.7	14	248	120	14.3
29.7	15	250	121	14

Properties of Saturated Steam			
Psig	Sensible Heat	Latent Heat	Total Heat
0	180	970	1150
1	183	968	1151
2	187	966	1153
3	190	964	1154
4	192	962	1154
5	195	960	1155
6	198	959	1157
7	200	957	1157
8	201	956	1157
9	205	954	1159
10	207	953	1160
11	209	951	1160
12	212	949	1161
13	214	948	1162
14	216	947	1163
15	218	945	1163

Boiler HP			
HP	Boiler Output Btuh	Boiler Input @ 80% Efficient	Pounds of Steam/ Hr
5	167,377	209,221	173
10	334,750	418,438	345
15	502,131	627,665	518
20	669,507	836,884	690
25	836,884	1,046,105	863
30	1,004,261	1,253,326	1,035
35	1,171,638	1,464,548	1,208
40	1,339,014	1,673,768	1,380
50	1,673,768	2,092,210	1,750
60	2,008,522	2,510,653	2100
70	2,343,275	2,929,094	2450
80	2,678,029	3,347,536	2800
90	3,012,782	3,765,978	3180
100	3,347,536	4,184,420	3500
125	4,184,420	5,230,525	4313
150	5,021,304	6,276,630	5175
175	5,858188	7,322,735	6038
200	6,695,072	8,368,840	6900
250	8,368840	10,461,050	8625
300	10,042,608	12,553,260	10350
350	11,716,376	14,645,470	12075

Steam Load Calculations	
For water -	Pounds of Steam/Hr = GPM/2 x Temperature rise
For fuel oil -	Pounds of Steam/Hr = GPM/4 x Temperature rise
For air -	Pounds of Steam/Hr = CFM/900 x Temperature rise
For radiation -	Pounds of Steam/Hr = EDR/4
Lifting condensate	Two feet of lift for each Psi
Calculate Sensible Heat	1.08 x CFM x Temp Rise or Delta T
Average heating load	25-40 Btuh/ square foot

Steam and Condensate Rules of Thumb

1 Btu	will raise 1 cu ft of air 55^{0}F
	will raise 55 cubic feet of air 1^{0}F
	Heat required to raise one pound of water one degree.
1 EDR	240 Btuh Steam
	Boiler Hp x 139
	0.25 Pounds Steam condensate/hr
	Btu/Hr /240 Steam
EDR Sq Ft.	Boiler Hp x 139
	Btu/Hr /240
	Lbs/Steam/ Hr /500
	Lb/Hr Condensate x 4
Lbs. Steam / Hr	Boiler Hp x 34.5
	Btuh divided by 960
One horsepower	0.746 KW
	746 WATTS
	2,545 BTUH
	1.0KVA
A boiler horsepower	10-11 ½ square feet of boiler heating surface.
	34,500 Btuh
	34.5 Lbs Steam/ Hr
	34.5 Lb H2O/ Hr
	0.069 GPM
	4.14 GPH
	140 EDR
1 Watt	3.415 BTUH
One kilowatt	1,000 WATTS
	3,413 BTUH
	1.341 HP
Lbs. Steam / Hr	Boiler Hp x 34.5
	Btuh divided by 960
	Btu/Hr dived by 240
Mbtu / Hr Output	Boiler Hp x 33.4
Evaporation Rate in GPM	Boiler Hp x .069
	½ GPM per 240,000 Btuh
	EDR/1000 x 0.5
	Lbs/Steam/ Hr /500
Condensate/ hr lb.	EDR divided by 4
Boiler Feedwater Rate	1 GPM per 240,000 Btuh Input
Factor of Evaporation	Sensible Heat + Latent Heat/970.3
Lb Steam condensate/ hr	EDR divided by 4

Flash Steam %	Temp of steam – temperature of condensate /970.4

Volume of Steam in Cubic Feet per Hour

High Pressure Steam					
PSIG Steam	Cu Ft/Lb.	PSIG Steam	Cu Ft/Lb.	**PSIG Steam**	Cu Ft/Lb.
25	11	100	3.9	**200**	2.18
50	7	125	3.23	**250**	1.74
75	5	150	2.76	**300**	1.47

Various Boiling Temperatures

Altitude		Pressure	
Altitude In feet	Boiling Temperature	Pressure	Boiling Temperature
-500	212.8^{0}F	29" Hg	76.6^{0}F
Sea Level	212^{0}F	25" Hg	133.2^{0}F
500	211^{0}F	20" Hg	161.9^{0}F
1,000	210^{0}F	2 Psi	218.6^{0}F
5,000	202.7^{0}F	5 Psi	227^{0}F
		10 Psi	239.3^{0}F
		15 psi	249.7^{0}F
		100 Psi	337.9^{0}F

Percentage of Flash Steam

Steam Psig	Flash Steam %	Steam Psig	Flash Steam %
0	0	8	3.31
2	.67	10	4.19
4	1.55	12	5.07
6	2.43	14	5.95

A 1/25" air curtain has the same R value as a four foot thick wall of iron.

How Air in System Affects Steam Temperature

Steam Psig	No Air	5% Air	10% Air	15% Air
2 Psig	219°F	216°F	213°F	210°F
5 Psig	227°F	225°F	222°F	219°F
10 Psig	239°F	237°F	233°F	230°F
20 Psig	259°F	256°F	252°F	249°F
50 psig	298°F	294°F	291°F	287°F
75 Psig	320°F	316°F	313°F	309°F
100 Psig	338°F	334°F	330°F	326°F

Air vent operating & drop away pressures

Hoffman

Model	Max Psig	Drop away Psig
1A	10	1 1/2
70A	15	11
40	10	6
1B	10	1 1/2
41	5	6
43	10	6
45	10	6
71A	15	11
71B	15	11
71C	15	11
4A	10	2
75	15	3
75H	15	10
76	15	3
3	25	
74	35	35
4	25	25
8C	125	

Watts

Model	Max Psig	Drop away Psig
SVA	10	1 1/2
SV	10	6
SVS	10	6

Maid o Mist

Model	Max Psig	Drop away Psig
All	10	6

Air capacity of piping per cubic foot

Pipe Size In	Length Ft		
	25	50	100
1 1/2	0.36	0.71	1.42
2	0.58	1.17	2.33
2 1/2	0.83	1.66	3.32
4	2.21	4.42	8.84
6	5.02	10.04	20.08
8	8.87	17.73	35.46

Pipe Size In	Length Ft		
	150	200	250
1 1/2	2.13	2.84	3.55
2	3.50	4.66	5.83
2 1/2	4.98	6.64	8.30
4	13.26	17.68	22.10
6	30.12	40.16	50.20
8	53.19	70.92	88.65

> The first steam heating system was installed in England so the governor of the Bank of England could grow grapes in cold weather.

Steam System Water Treatment

Water Treatment Formulas	
Concentration cycles	PPM chlorides in boiler water / ppm feedwater
Test for carryover	Conductivity over 25 µmhos in condensate water
	Boiler treatment chemicals in the condensate.

Typical Water Quality Parameters in Steam Systems		
Feedwater	Softness	Less than 1 ppm
Boiler Water	Hardness	Less than 1 ppm
	Ph	7-9 or 9.5-11*
	TDS	1,500-3000 ppm
		2,000-4000 µS
	Sulfite	30-60 ppm
	Hydroxyl Alkalinity	200 – 400 ppm
Condensate	pH	8.2-9.0
	TDS	Less than 20 ppm
*Verify with boiler manufacturer		

Common Causes of Carryover in Steam System	
Mechanical	**Chemical**
Boiler design	Foaming
Undersized boiler	Suspended / dissolved solids
Improperly piped zone valves	Organic contamination
Pipe insulation removed	High alkalinity
Excessive cold starting	Process contamination
Quick opening control valves	Pipe threading oil
Vacuum forming in system	Incorrect cleaning of boiler
	Excessive chemical treatment
	Silica

Calculate Condensate Return Percentage
$1 - \dfrac{\textit{Feedwater Chloride Level}}{\textit{Makeup Chloride Level}}$
$1 - \dfrac{\textit{Feedwater Silica Level}}{\textit{Makeup Silica Level}}$
$1 - \dfrac{\textit{Feedwater Conductivity Level}}{\textit{Makeup Conductivity Level}}$

O2 % in Water Based by Temperature					
^{0}F	^{0}C	O2 PPM	^{0}F	^{0}C	O2 PPM
50	10	11.1	140	60	4.7
60	15.6	10	150	65.6	4.3
70	21	9.0	160	71.1	3.9
80	27	8.2	170	76.6	3.4
90	32	7.5	180	82.2	2.7
100	37.8	6.9	190	87.8	2
110	43.3	6.3	200	93.3	1.3
120	48.9	5.7	210	98.9	0.6
130	54.4	5.2			

To get 1,000,000 Btu's you need the following:
1,000 Cu ft of natural gas
10 Therms of natural gas
1 Mcf or dekatherm of natural gas
10.92 gallons of propane
8 gallons of gasoline
7.19 gallons of #2 fuel oil
293.083 Kwh of electricity
29.3 boiler horsepower

Never plug the vent on a condensate or boiler feed tank.

Sizing a Steam Boiler

Steam boilers are sized according to the connected load plus an allowance for piping. Once you add all the heat emitters, add 20% for the piping.

Estimated Btu's per section of steam radiator. Thin Tube radiators

Radiator Ht In.	Number of Tubes				
	3	4	5	6	7
13"					630
16 ½"					840
18"					
20"	420	540	638	720	1,020
22"					
23"	480	600	720	840	
26"	506	660	840	960	
30"	720				
32"		840	1,039	1,200	
36"	840				
37"		990	1,200	1,440	

2 psi steam @ 70°F room temperature.

Estimated Btu's per section of steam radiators. Cast Iron Radiators

Radiator Ht In.	Number of Columns			
	1	2	3	4
18"			540	720
20"	360	480		
22"			720	960
23"	398	559		
26"	480	638	900	1,200
30"				
32"	600	799	1,080	1,560
36"				
37"				
38"	720	960	1,200	1,920
45"		1,200	1,440	2,400

2 psi steam @ 70°F room temperature.

Pipe Insulation

Heat Loss from Insulated Pipe

Pipe Size Inches	Insulation Thickness	Btus lost*
½"	1"	0.157
¾"	1"	0.177
1"	1"	0.200
1"	1 ½"	0.167
1 ¼"	1"	0.282
1 ¼"	1 ½"	0.193
1 ½"	1"	0.255
1 ½"	1 ½"	0.210
2"	1"	0.297
2"	1 ½"	0.240
2 ½"	1"	0.340
2 ½"	1 ½"	0.270
3"	1"	0.395
3"	1 ½"	0.312
4"	1"	0.480
4"	1 ½"	0.379

Based on 225°F Internal Pipe Temperature

Heat Loss from Uninsulated Horizontal Pipe
Btuh / linear ft. @ 70°F room temperature

Nom Pipe Size	Steam @ 5 PSIG
½	96
¾	118
1	144
1 ¼	170
1 ½	202
2	248
2 ½	296
4	448

Never plug the vent on a condensate or boiler feed tank.

Condensate Return

Condensate Pump Sizing Rules of Thumb

Pump GPM =	Evaporation Rate x 3
Tank Sizing =	Pump GPM x 1
Evap rate = Boiler HP x .069 or Lbs Steam/Hr / 500	

Boiler Feed System Sizing

Pump GPM =	Evaporation Rate x 2
Tank Sizing =	Pump GPM x 20*
Based on 20 minutes storage. 1 gal storage per Boiler HP	

Condensate Tank & Pump

Boiler HP	Evaporation Rate	Pump GPM	Tank Size Gallons
5	0.345	1.035	1.035
10	0.69	2.07	2.07
15	1.035	3.105	3.105
20	1.38	4.14	4.14
25	1.725	5.175	5.175
30	2.07	6.21	6.21
35	2.415	7.25	7.25
40	2.76	8.28	8.28
45	3.105	9.315	9.315
50	3.45	10.35	10.35

Boiler Feed Tank Sizing

Boiler HP	10 Minutes	20 Minutes	30 Minutes	Pump GPM
20	21	41	62	2.07
30	31	62	93	3.10
40	41	83	124	4.14
50	52	103	155	5.17
60	62	124	186	6.21
70	72	145	217	7.24
80	83	166	248	8.28
90	93	186	279	9.31
100	103	207	310	10.35
110	114	228	341	11.38
120	124	248	372	12.41
130	134	269	403	13.45
140	145	290	435	14.48
150	155	310	466	15.52
160	166	331	497	16.55
170	176	352	528	17.59
180	186	372	559	18.62
Tank & pump sizing based upon 50% safety factor				

Steam & Condensate Piping

One Pipe System – Steam Pipe Sizing
Parallel flow with dry return

Pipe Size In.	Btuh Net	Pounds / Hr
2"	92,640	97
2 ½"	152,400	159
4"	589,680	614
6"	1,834,000	1,911

Dry return sizing

Pipe Size In.	Btuh Net	Pounds / Hr
1"	76,800	80
1 ¼"	161,280	168
1 ½"	254,400	265
2"	552,000	575

One Pipe System – Pipe Sizing
Parallel flow with wet return

Pipe Size In.	Btuh Net	Pounds / Hr
2"	92,640	97
2 ½"	152,400	159
4"	589,680	614
6"	1,834,080	1,911

Wet return sizing

Pipe Size In.	Btuh Net	Pounds / Hr
1"	168,000	175
1 ¼"	288,000	300
1 ½"	456,000	475
2"	960,000	1,000

Boiler Equalizer Connection

Pipe Size In.	Btuh Net	Pounds / Hr
1 ½"	216,000	225
2 ½"	1,536,000	1,600
Source: The Color of Steam Peerless		

Boiler Equalizer Connection

Pipe Size In.	Btuh Net	Pounds / Hr
1 ½"	216,000	225
2 ½"	1,536,000	1,600
Source: The Color of Steam Peerless		

A 1" pipe can transport as many Btus as a 20" round duct.

Two Pipe Steam Pipe Capacities
When condensate flows in same direction as steam

Pipe Size In	100 feet		200 Feet	
	EDR	Btu's	EDR	Btu's
2	650	156,000	461	110,640
2 1/2	1,030	247,200	731	175,440
4	3,800	912,000	2,698	647,520
6	11,250	2,700,000	7,987	1,916,880
8	22,250	5,340,000	15,797	3,791,280
10	40,800	9,792,000	28.968	6,952,320
12	66,000	158,40,000	46,860	11,246,400

Condensate Return Main Capacity Horizontal Pitch
2 Ounce pressure drop per 100 inches

Pipe Size	Dry Return		Wet Return	
	EDR	Btu's	EDR	Btu's
1"	370	88,800	900	216,000
1 1/4"	780	187,200	1,530	367,200
1 1/2"	1,220	292,800	2,430	583,200
2"	2,660	638,400	5040	1,209,600
2 1/2"	4,420	1,060,800	8,460	2,030,400
4"	17,380	4,171,200	27,900	6,696,000

Condensate piping should be schedule 80

Flow Rate in Pounds per Hour
Steam Flow in Schedule 40 Pipe @ 3.5 Psig

Pipe Size	Pressure Drop per 100 Feet in ounces			
	1	2	4	8
3/4"	9	14	20	29
1"	17	26	37	54
1 1/4"	36	53	78	111
1 1/2"	56	84	120	174
2"	108	162	234	336
2 1/2"	174	258	378	540
4"	640	950	1,410	1,960
6"	1,920	2,820	3,960	5,700
8"	3,900	5,570	8,100	11,400
10"	7,200	10,200	15,000	21,000
12"	11,400	16,500	23,400	33,000

Typical pressure drop for residential systems is 1 ounce per 100 feet

Typical pressure drop for commercial systems is 2 ounces per 100 feet

Standard Nipples & Pipe Sizing Schedule 40

Pipe Size	Outside Diameter (O.D.)	Circumference
1/8"	0.405"	1.272"
1/4"	0.540"	1.696"
3/8"	0.675"	2.121"
1/2"	0.840'	2.639"
3/4"	1.050"	3.299"
1"	1.315"	4.131"
1 1/4"	1.660"	5.215"
1 1/2"	1.900"	5.969"
2"	2.375"	7.461"
2 1/2"	2.875"	9.032"
3"	3.500"	10.995"
4"	4.500"	14.137"
5"	5.563"	17.476"
6"	6.625"	20.812"
8"	8.625"	27.095"

Equivalent Length
Length of pipe to be added for each fitting

Pipe size inches	Standard Elbow	Gate Valve	Globe valve
1/2	1.3	0.3	14
3/4	1.8	0.4	18
1	2.2	0.5	23
1 1/4	3.0	0.6	29
1 1/2	3.5	0.8	34
2	4.3	1.0	46
2 1/2	5.0	1.1	54
4	9	1.9	92
6	13	2.8	136
8	17	3.7	180

When checking a pop safety valve, the boiler pressure should at 75% of the safety valve setting.

Steam Velocity

Verify steam velocity the manufacturer recommends. I like using 40 feet per second velocity in near boiler piping and 50 feet per second for system piping.

Pipe Sizing for Steam Velocities Below 40 FPS

Pipe Size	Lbs / Hr	Btu/ Hr
2"	140	134,400
2 ½"	199	191,200
4"	530	509,200
6"	1,204	1,155,600
8"	2,131	2,046,000
10"	3,397	3,262,000
12"	4,783	4,592,000

Pipe Sizing for Steam Velocities Below 50 FPS

Pipe Size	Lbs / Hr	Btu/ Hr
2"	140	134,400
2 ½"	199	191,200
4"	530	509,200
6"	1,204	1,155,600
8"	2,131	2,046,000
10"	3,397	3,262,000
12"	4,783	4,592,000

Feet per Second to Miles per Hour

FPS	MPH	FPS	MPH	FPS	MPH
15	10.2	55	37.5	110	75
20	13.6	60	40.9	120	81.8
25	17	65	44.3	130	88.6
30	20.4	70	47.7	140	95.5
35	23.9	75	51.1	150	102.3
40	27.2	80	54.5	160	109
45	30.7	90	61.4	170	115.9
50	34	100	68.1	180	122.7

How to calculate steam velocity

Lbs./Hr x Cubic Volume of Steam divided by 25 x Internal area of pipe = Steam Velocity

$$Velocity = \frac{Lbs\ Hour \times Cubic\ Volume\ of\ Steam}{25 \times Internal\ Area\ of\ Pipe}$$

Lbs. Steam/Hr = Btuh/960
Lbs. Steam/Hr = Boiler HP x 34.5

Internal Volume of Schedule 40 Pipe

Pipe Size	Internal Square Inches	Pipe Size	Internal Square Inches	Pipe Size	Internal Square Inches
2"	3.36"	4"	12.73"	8"	51.15"
2 ½"	4.78"	5"	19.99"	10"	81.55"
3"	7.39"	6"	28.89"	12"	114.80"

Velocity

Knowing	Multiply by	To Get
Ft/Sec	60	Ft/Min
Ft/min	0.01139	Miles/Hr
	0.01667	Ft/Second
Cu Ft/ Min	0.1247	Gal/Sec
Cu Ft/Sec	448.8	Gal/min
Miles/Hr	88	Ft /Min
Gal/minute	0.002228	Cu Ft/Sec

One cubic inch of water at atmospheric pressure will become one cubic foot of steam when boiled.

Combustion Air

Combustion Air Openings

One Opening

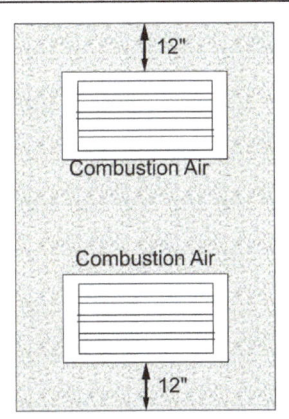

Two Openings

Mechanical Combustion Air
0.35 CFM per 1,000 Btuh Boiler Input
Indoor Air
50 cubic feet of volume for each 1,000 Btuh

Typical Free Area Estimate for Various Openings and Louvers	
Opening Type	Estimated A_K or Free Area
Metal Louver	75% Free Area
Wooden Louver	50% Free Area
Metal Mesh Screen	98% Free Area
Motorized Combustion Air Dampers	95% Free Area
A_K or Area Factor is the actual free area of the grill.	

Btuh Capacity of Direct Vent Metal Combustion Air Louvers			
Height Inches	Btu Capacity		
	Width in inches		
	12	24	36
12	432,000	864,000	1,296,000
18	648,000	1,296,000	1,944,000
24	864,000	1,728,000	2,592,000
30	1,080,000	2,160,000	3,240,000
36	1,296,000	2,592,000	3,888,000
42	1,512,000	3,024,000	4,536,000
48	1,728,000	3,456,000	5,184,000
60	2,160,000	4,320,000	6,480,000

Based on direct connect metal louvers with 75% free area and 4,000 Btuh per inch of free space.

Two Openings

Two Openings
One opening within a foot of the floor and one within a foot of the ceiling.
One inch of free area for every 4,000 Btu input of the boiler for a direct opening or a vertical duct.
One inch of free area for every 4,000 Btu input of the boiler for a direct opening or a vertical duct.
One inch of free area for every 2,000 Btu input of the boiler for a horizontal duct.

Water transfers heat 24 times faster than air.

One Opening
The opening should be within a foot of the ceiling.
One inch of free area for every 3,000 Btu input for a direct opening or a vertical duct.

Volume of Air			
Degrees F	Cubic ft in 1 lb.	Degrees F	Cubic ft in 1 lb.
0	11.58	130	14.85
32	12.38	140	15.1
40	12.58	150	15.35
50	12.84	160	15.60
62	13.14	170	15.85
70	13.34	180	16.1
80	13.59	200	16.6
90	13.85	210	16.86
100	14.09	212	16.91
110	13.34	220	17.11
120	14.59		

Mechanical Combustion Air
Based on the International Mechanical Code of 0.35 CFM per 1,000 Btuh.

Btuh	CFM	Btuh	CFM
50,000	18	750,000	263
75,000	26	800,000	280
100,000	35	850,000	298
150,000	53	900,000	315
200,000	70	950,000	333
250,000	88	1,000,000	350
300,000	105	1,500,000	525
350,000	123	2,000,000	700
400,000	140	2,500,000	875
450,000	158	3,000,000	1,050
500,000	175	3,500,000	1,225
550,000	193	4,000,000	1,400
600,000	210	4,500,000	1,575
650,000	228	5,000,000	1,750
700,000	245		

Ducted Combustion Air to Burner
Power Flame recommends sizing the combustion air duct for a pressure drop of 0.1" w.c. including all screens, filters, and fittings.

Direct Vented Combustion Air Sizing
Estimated Combustion Air Required @ Various Boiler Inputs

Boiler Btuh Input	Cu Ft Gas / Hr**	Cu Ft Gas/ Minute	CFM Air*
200,000	200	3.33	50
300,000	300	5.0	75
400,000	400	6.67	100
500,000	500	8.33	125
600,000	600	10.0	150
700,000	700	11.67	175
800,000	800	13.33	200
900,000	900	15.0	225
1,000,000	1,000	16.67	250
1,500,000	1,500	25.0	375
2,000,000	2,000	33.33	500
3,000,000	3,000	50.0	750

* Based upon 15 cubic feet of air for every cubic foot of gas burned.

** Based on 1,000 Btu per cubic foot of gas

Excess Air

Parts of Air per One Part of Gas	Excess Air Percentage
10	0%
11	10%
12	20%
13	30%
14	40%
15	50%
16	60%
17	70%
18	80%
19	90%
20	100%

According to the EPA, a boiler is most efficient when operated between 50-80% of the rated capacity.

Natural Gas /Fuel Oil

Btuh	Cu Feet/Hr	Cubic Feet/ Min	Cubic Feet/ Sec
		Boiler Gas Consumption	
200,000	200	3.33	0.05
400,000	400	6.66	0.11
600,000	600	10	0.17
800,000	800	13.3	0.22
1,000,000	1,000	16.67	0.28
2,000,000	2,000	33.33	0.56
3,000,000	3,000	50.00	0.83
4,000,000	4,000	66.67	1.11
5,000,000	5,000	83.33	1.39
Based upon 1,000 Btu's per cubic foot			

Gas pipe sizing

Steel Pipe Size	10 Feet	20 Feet	40 feet	80 Feet
		Pipe Length		
	Capacity in Cubic Feet per hour			
½"	120	85	60	42
¾"	272	192	136	96
1"	547	387	273	193
11/4"	1,200	849	600	424
1 ½"	1,860	1,316	930	658
2"	3,759	2,658	1,880	1,330
2 ½"	6,169	4,362	3,084	2,189
4"	23,479	16,602	11,740	8,301

Maxitrol RV Series Gas Pressure Regulator
Spring Color Ratings

Spring Color	Inches W. C.	Spring Color	Inches W. C.
Plated	3-6"	Violet	4-12"
Orange	4-8"	Green	5-15"
Blue	5-12"	Red	10-22"
Brown	1-3.5"	Yellow	15-30"
Plated	2-5"	Black	20-42"
Pink	3-8"		

Gas Pressure Comparison

Inches Hg	Ounces	PSI
0.1	0.05	0.003
0.2	0.12	0.007
0.4	0.23	0.01
0.6	0.35	0.02
0.8	0.46	0.028
1	0.58	0.036
2	1.15	0.072
3	1.73	0.108
4	2.31	0.144
5	2.89	0.18
6	3.46	0.216
7	4.04	0.252
8	4.62	0.288
9	5.20	0.324
10	5.78	0.36
11	6.35	0.396
12	6.93	0.432
13	7.51	0.468
14	8.09	0.504
15	8.67	0.54
16	9.24	0.576
17	9.82	0.612
18	10.4	0.648
19	10.98	0.684
20	11.56	0.72
21	12.13	0.756
22	12.71	0.792
23	13.29	0.828
24	13.87	0.864
25	14.45	0.9
26	15.02	0.936
27	15.60	0.972
28	16.18	1.008

The relief valve was patented in 1682 by Denis Papin

Natural Gas Calculations	
1 Cu ft Natural Gas =	1,000 Btus
1 MCF =	1,000,000 Btus
	1 MMBTU
	1 MCF
	1,000 Cu Ft.
	10 CCF
	10 Therms
1 Dekatherm	1 MCF
	10 Therms
	1,000,000 Btus
1 Therm =	100,000 Btus or 100 MBTU
	0.1 MCF
	1 CCF
	100 Cu. Feet
1 CCF =	1,000 Cu Ft
	100 Therm
Propane	
1 gallon =	92,000 Btus
1 Cu Foot =	2,250 Btus
#2 Fuel Oil	
1 Gallon =	140,000 Btus
Petroleum	
1 Barrel =	42 gallons

Leak testing gas valves.

The electric gas valves for the boiler should be tested yearly. Code allows a certain amount of acceptable leakage through a closed gas valve. The following are the recommendations provided by Honeywell for testing their gas valves.

How to leak test a gas valve

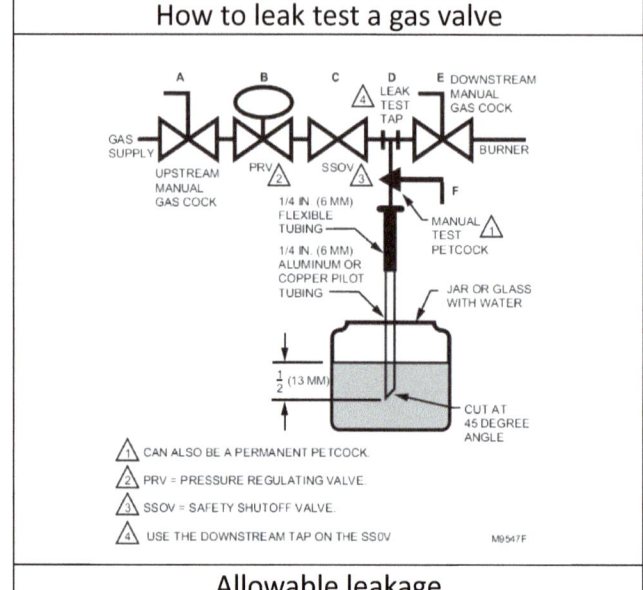

Allowable leakage

Table 5. Allowable Leakage for V48/V88 Valves.

V48/V88 Pipe Size (in.)	Air (cc/h)	Natural Gas (cc/h)[a]	Bubbles/10 sec.; Max @ 45 degrees[b]
3/4	266	332.5	8
1	302	377.5	9
1-1/4	442	552.5	13
1-1/2	442	552.5	13
2	650	812.5	20
2-1/2	650	812.5	20
3	650	812.5	20

[a] Natural gas: multiply air by 1.25.
[b] Bubble leakage: Divide natural gas by 573, then multiply by 14.

Leak test fitting

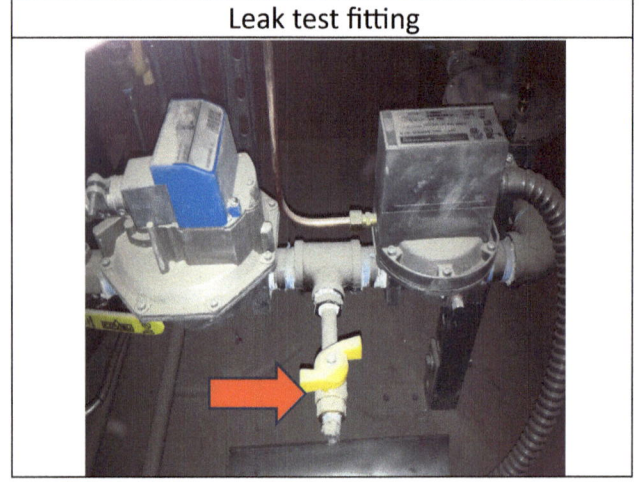

Oxygen corrosion is twice as corrosive @ 86°F than it is @122°F in water.

#2 Fuel Oil

#2 Fuel Oil Delivery Rate in Gallons per Hour

Nozzle Rating	Pump Pressure				
	100	125	140	150	175
1.0	1.0	1.1	1.18	1.2	1.3
1.5	1.5	1.7	1.77	1.8	2.0
2.0	2.0	2.2	2.37	2.5	2.6
2.5	2.5	2.8	2.96	3.1	3.3
3.0	3.0	3.3	3.55	3.7	3.9
3.5	3.5	3.9	4.14	4.3	4.6
4.0	4.0	4.5	4.73	4.9	5.3
4.5	4.5	5.0	5.32	5.5	5.9
5.0	5.0	5.6	5.95	6.1	6.6
5.5	5.5	6.2	6.5	6.7	7.3
6.0	6.0	6.7	7.1	7.3	7.9
7.0	7.0	7.8	8.28	8.5	9.2
8.0	8.0	8.9	9.47	9.8	10.5
9.0	9.0	10.0	10.7	11.0	11.8
10.0	10.0	11.2	11.8	12.2	13.2
11.0	11.0	12.3	13.0	13.4	14.5
12.0	12.0	13.4	14.2	14.6	15.8
13.0	13.0	14.5	15.4	16.0	17.2
14.0	14.0	15.5	16.6	17.0	18.4
15.0	15.0	16.8	17.7	18.4	19.8

Nozzle Rating	Pump Pressure				
	200	225	250	275	300
1.0	1.4	1.5	1.6	1.7	1.75
1.5	2.1	2.3	2.4	2.5	2.6
2.0	2.9	3.0	3.2	3.3	3.5
2.5	3.5	3.8	4.0	4.1	4.3
3.0	4.2	4.5	4.7	5.0	5.2
3.5	5.0	5.3	5.5	5.8	6.0
4.0	5.6	6.0	6.3	6.6	6.9
4.5	6.3	6.8	7.1	7.4	7.8
5.0	7.1	7.5	7.9	8.3	8.7
5.5	7.8	8.3	8.7	9.1	9.5
6.0	8.5	9.0	9.5	9.9	10.4
7.0	9.9	10.5	11.0	11.6	12.0
8.0	11.3	12.0	12.6	13.2	13.8
9.0	12.6	13.5	14.2	14.8	15.6
10.0	14.1	15.0	15.8	16.6	17.3
11.0	15.0	15.6	15.8	16.6	17.3
12.0	16.9	18.0	18.9	19.8	20.7
13.0	18.4	19.5	20.6	21.5	22.5
14.0	19.8	21.0	22.0	23.1	24.0
15.0	21.3	22.5	23.6	24.8	25.9

Water Equations

Water Conversion Factors

Knowing	Multiply by	Desired result
US Gallons	8.34	Pounds
	0.1338	Cubic Feet
	0.00379	Cubic Meters
	231	Cubic Inches
	3.7853	Liters
Cu Inch water	0.03613	Pounds
	0.004329	US Gallons
	0.576384	Ounces
Pounds Water	27.72	Cu Inches
	0.12	US Gallons
PSIG	2.307	Height of water in feet
Lbs. Water	0.11992	Gallons
	0.45419	Liters
	27.643	Cu. Inches
	0.01603	Cu. Feet
	0.000454	Cu. Meters

Water Conversion Factors

1 cubic foot of water	7.5 gallons
	1,728 Cubic Inches
	62.4 Pounds
Estimate Btus	500 x GPM x Delta T
Delta T	BTU /500 x GPM
GPM =	BTU / 500 x Delta T
1 Pd of Water	27.72 Cu Inches@65 Deg F
Pump HP	(GPM x Total head in feet) / 3960
GPM flow rate of water through a pipe	0.0408 x (pipe diameter)^2 x (water velocity)
Max water velocity	< 6 FPS @ 200^{0}F
1 Teaspoon	60 drops
1 Tablespoon	3 teaspoons
1 Ounce	2 tablespoons
1 Pint	2 cups
1 Quart	2 pints
1 Gallon	4 quarts
1 Measuring cup	16 tablespoons
	8 ounces

Electrical

It's a good idea to check the tightness of electrical wires on a boiler. They loosen over time and could cause intermittent failures.

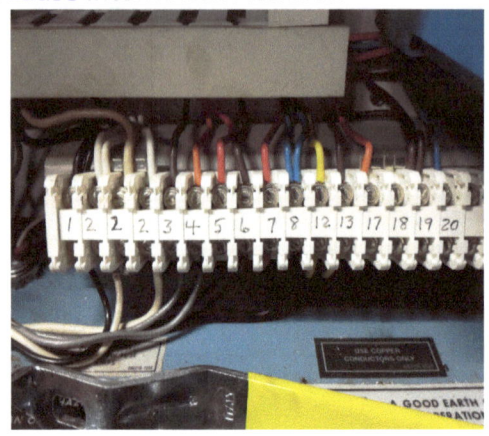

Sizing an Electric Boiler

KW/pd /Steam based on feedwater temperatures (^{0}F)			
	Steam Pressure Psig		
Feedwater Temperature	2	10	15
40	.3355	.3375	.3388
60	.3296	.3316	.3329
80	.3238	.3278	.3271
100	.3179	.3199	.3212
120	.3210	.314	.3154
140	.3062	.3082	.3095
160	.3003	.3029	.3036
180	.2945	.2964	.2978
200	.2886	.2906	.2919

Electrical Equivalents Formulas

Watt=	44.236 foot-pounds minute
	2,654.16 foot-pounds hour
	0.00134 hp
	3.414 Btu
	0.0035 lb. of water evaporated per hour
	44.236 foot-pounds minute
	2,654.16 foot-pounds hour
Kilowatt=	44,235 foot-pounds minute
	1.34 H.P.
	0.955 BTU per second
	57.3 Btu per minute
	3,438 Btu per hour
	1,000 W
	1.34 hp

	3.53 lbs. water evaporated per hr from and at 212 degrees F
	0.955 Btu's
	57.3 BTU per minute
	3,413 Btuh per hour
1 H.P.=	33,000 foot-pounds minute
	746 watts
	42.746 Btu per minute
	2,564.76Btu per hour
1 Btu=	772 ft lbs.
	17.452 watts per minute
	0.2909 watts hour

Fuse Sizing Guide

Class CC Time Delay

Multiply Motor FLA from nameplate x 2. Round up or down to nearest fuse size
Example:
1 HP @ 115 Volt = 15.1 Full Load Amps
15.1 x 2 = 30.2 Use 30 Amp fuse

Class J Time Delay

Multiply Motor FLA from nameplate x 1.5. Round up or down to nearest fuse size.
Example:
½ HP @ 115 Volt = 7.4 Full Load Amps
7.4 x 1.5 = 11.10 Use 12 Amp fuse

Standard 24 Volt Thermostat Connections

Terminal	Usage	Wire colors
R or V	24 V power	Red
Rh or 4	24 V Heating Power	Red
Rc	24 V Cooling Power	Red
C	24 V Common	Black
Y	1st Stage Cooling	Yellow
Y2	2nd Stage Cooling	Blue or Orange
W	1st Stage Heat	White
W2	2nd Stage Heat	No Standard Color
G	Fan	Green

A condensate pipe is usually one size smaller than the steam pipe feeding it.

Wire Size & Amp Ratings Copper			
	140⁰F	167⁰F	194⁰F
Wire Types	NM-B, UF-B	THW, THWN, SE, USE, XHHW	THWN-2, THHN, XHHW-2
Wire Gauge	Amp ratings		
14	15	20	25
12	20	25	30
10	30	35	40
8	40	50	55
6	55	65	75
4	70	85	95
3	85	100	115
2	95	115	130
1		130	145
1/0		150	170
2/0		175	195
3/0		200	225
4/0		230	260

Full Load Amperes of Single-Phase Motors			
HP	RPM	115V	230V
1/8	1725	2.8	1.4
	1140	3.4	1.7
	860	4.0	2.0
1/4	1725	4.6	2.3
	1140	6.15	3.07
	860	7.5	3.75
1/3	1725	5.2	2.6
	1140	6.25	3.13
	860	7.35	3.67
1/2	1725	7.4	3.7
	1140	9.15	4.57
	860	12.8	6.4
3/4	1725	10.2	5.1
	1140	12.5	6.25
	860	15.1	7.55
1	1725	13.0	6.5
	1140	15.1	7.55
	860	15.9	7.95

Full Load Amperes of Three Phase Motors			
HP	RPM	115V	230V
¼	1725	0.95	0.48
	1140	1.4	0.7
	860	1.6	0.8
1/3	1725	1.19	0.6
	1140	1.59	0.8
	860	1.8	0.9
½	1725	1.72	0.86
	1140	2.15	1.08
	860	2.38	1.19
¾	1725	2.46	1.23
	1140	2.92	1.46
	860	3.26	1.63
1	1725	3.19	1.6
	1140	3.7	1.85
	860	4.12	2.06
1 ½	1725	4.61	2.31
	1140	5.18	2.59
	860	5.75	2.88
2	1725	5.98	2.99
	1140	6.5	3.25
	860	7.28	3.64
3	1725	8.70	4.35
	1140	9.25	4.62
	860	10.3	5.15
5	1725	14.0	7.0
	1140	14.6	7.3
	860	16.2	8.1
7 ½	1725	20.3	10.2
	1140	20.9	10.5
	860	23.0	11.5

Color Code of NB-B Romex Cable	
The following is the color code of the cable so you can quickly tell the gauge of the wire.	
Jacket Color	Wire Size
White	14 AWG
Yellow	12 AWG
Orange	10 AWG
Black	8 AWG
Black	6 AWG

Venting the Boilers

7 Times Rule - The flow area of the largest common vent or stack shall not exceed seven times the area of the smallest draft hood outlet.

Horizontal vs. Vertical - The horizontal vent (L) must be no more than 75% of the vertical height (H) of the flue. If using B Vent, the horizontal length can be the same as the height of the chimney, as per International Fuel Gas Code, 2006 503.10.9.

7 Times Rule Round Flue			
Smallest Draft Hood Outlet	Largest Common Flue	Smallest Draft Hood Outlet	Largest Common Flue
3"	7"	7"	18"
4"	10"	8"	22"
5"	12"	9"	22"
6"	14"	10"	26"

7 Times Rule Rectangular Flue		
Smallest Draft Hood Outlet	Area of outlet	Largest Common Vent Area
3"	7.06"	49"
4"	12.56"	88"
5"	19.64"	137"
6"	28.27"	198"
7"	38.48"	269"
8"	50.27"	352"
10"	78.54"	550"

Appliance Venting Categories			
Cat	Condensing	Efficiency	Pos/Neg
I	Non-condensing	83% or less	Negative
II	Condensing	Over 83%	Negative
III	Non-Condensing	83% or less	Positive
IV	Condensing	Over 83%	Positive

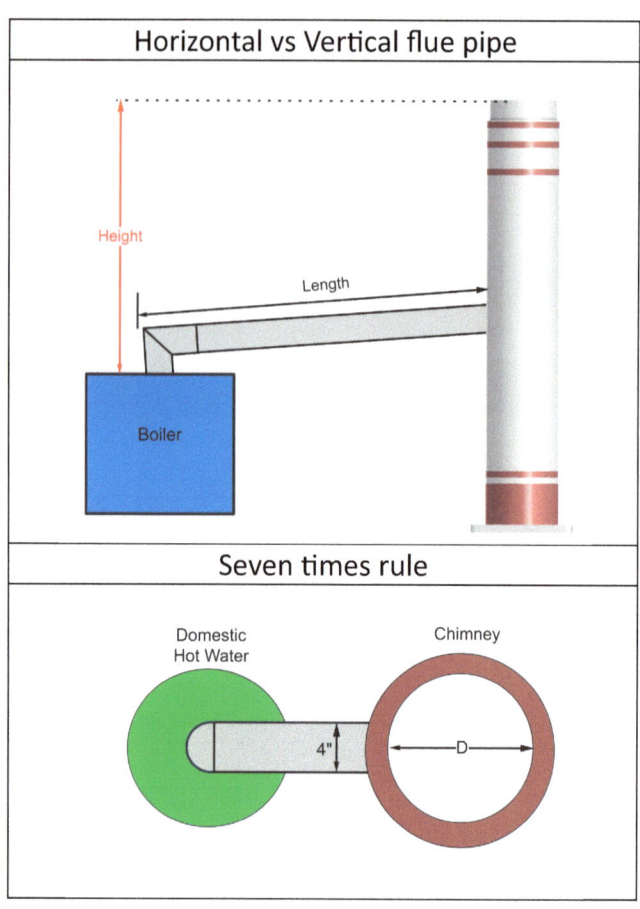

Horizontal vs Vertical flue pipe

Seven times rule

Sidewall Venting

If venting the boiler through the sidewall, the following are the International Mechanical Codes covering the installation:

Where adjacent to walkways, the termination of mechanical draft systems shall not be less than 7 feet above walkway.

3 feet above any forced air inlet within 10 feet

4 feet below, 4 feet horizontally from or 1 foot above any door, window, or gravity vent into building

No closer than 3 feet from an interior corner formed by 2 walls perpendicular to each other.

Not within 3 feet horizontally or directly above an oil tank or gas meter

At least 12 inches above finished grade.

Rectangular Storage Tank Capacity in Gallons

Length" x Width" x Height" divided by 231 = Gallons

Rectangular Tank Height 12"

US Gallons

Length	Width			
	12	18	24	36
12	7	11	15	22
18	11	17	22	34
24	15	22	30	45
30	19	28	37	56
36	22	34	45	67
42	26	39	52	79
48	30	45	60	90
60	37	56	75	112

Rectangular Tank Height 18"

Length	Width			
	12	18	24	36
12	11	17	22	34
18	17	25	34	50
24	22	34	45	67
30	28	42	56	84
36	34	50	67	101
42	39	59	79	118
48	45	67	90	135
60	56	84	112	168

Rectangular Tank Height 24"

Length	Width			
	12	18	24	36
12	15	22	30	45
18	22	34	45	67
24	30	45	60	90
30	37	56	75	112
36	45	67	90	135
42	52	79	105	157
48	60	90	120	180
60	75	112	150	224

Rectangular Tank Height 36"

Length	Width			
	12	18	24	36
12	22	34	45	67
18	34	50	67	101
24	45	67	90	135
30	56	84	112	168
36	67	101	135	202
42	79	118	157	236
48	90	135	180	269
60	112	168	224	337

Circular Storage Tank Capacity in Gallons

Multiply ½ Tank Diameter by itself.

Multiply that by 3.146 x Length of Tank in inches.

Divide by 231 = Gallons of water

Circular tanks

Estimate Storage Tank Capacity in Gallons

Length (feet)	US Gallons				
	18	24	30	36	42
1	1.1	1.96	3.06	4.41	5.99
2	26	47	73	105	144
2.5	33	59	91	131	180
3	40	71	100	158	216
3.5	46	83	129	184	252
4	53	95	147	210	288
4.5	59	107	165	238	324
5	66	119	181	264	360
5.5	73	130	201	290	396
6	79	141	219	315	432
6.5	88	155	236	340	468
7	92	165	255	368	504
7.5	99	179	278	396	540
8	106	190	291	423	576
9	119	212	330	476	648
10	132	236	366	529	720
12	157	282	440	634	864
14	185	329	514	740	1008

Length (feet)	US Gallons				
	Inside Diameter (Inches)				
	48	54	60	66	72
1	7.83	9.91	12.24	14.41	17.62
2	188	238	294	356	423
2.5	235	298	367	445	530
3	282	357	440	534	635
3.5	329	416	513	623	740
4	376	475	586	712	846
4.5	423	534	660	800	852
5	470	596	734	899	1057
5.5	517	655	808	978	1163
6	564	714	880	1066	1268
6.5	611	770	954	1156	1374
7	658	832	1028	1244	1480
7.5	705	889	1101	1355	1586
8	752	949	1175	1424	1691
9	846	1071	1322	1599	1903
10	940	1189	1463	1780	2114
12	1128	1428	1762	2133	2537
14	1316	1666	2056	2490	2960

Conversion Factors	
1 Mile =	1,760 yards
	5,280 feet
	63,360 inches
	1.609 Km
1 Foot =	0.3048 M
	30.48 Cm
	304.8 mm
1 Inch =	25,400 microns
1 acre =	43,560 Sq. Ft
	4,840 Sq. Yds
	0.4047 Hectares
1 Sq. Mile =	640 Acres
1 Sq. Yd =	9 Sq. Ft
	1,296 Sq. Inches
1 Sq. Foot =	144 Square Inches
1 Cu Yard =	27 Cu Ft
	46,656 cu inches
	1,616 pints
	807.9 quarts
	764.6 Liters
1 Cu foot =	1,728 cubic inches
1 Gallon H2O =	8.333 Lbs
Diameter of Circle =	Circumference x 0.3188
Circle Circumference =	Diameter x 3.1416
1 League =	3.0 Miles
1 gallon =	4 quarts
	8 pints
	3.785 liters
	0.13368 Cu Feet
	231 Cu Inches
1 Liter =	0.2642 Gallons
	1.057 quarts
	2,113 pints

Liquid Measurements	
1 Teaspoon =	60 drops
1 Tablespoon =	3 teaspoons
1 Ounce =	2 tablespoons
1 Measuring cup =	16 tablespoons or 8 ounces
1 Pint =	2 cups
1 Quart =	2 pints
1 Gallon =	4 quarts

Circular Equivalent of Rectangular Duct						
	Rectangular Height in Inches					
Rectangular Width	6	8	10	12	14	16
	Circular Equivalent					
6	7	8	8	9	10	10
8	8	9	10	11	11	12
10	8	10	11	12	13	14
12	9	11	12	13	14	15
14	10	11	13	14	15	16
16	10	12	14	15	16	17
18	11	13	15	16	17	19
20	11	13	15	17	18	20
22	12	14	16	18	19	20
24	12	15	17	18	20	21
26	13	15	17	19	21	22
28	13	16	18	20	21	23
30	14	16	18	20	22	24
36	15	17	20	22	24	26
42	16	19	21	23	26	28
48	17	20	22	25	27	29

Common Fraction to Decimal to Millimeters		
Fraction	Decimal	Millimeters
1/16	0.0625	1.587
1/8	0.125	3.175
3/16	0.1875	4.762
1/4	0.250	6.350
5/16	0.3125	7.937
3/8	0.375	9.525
7/16	0.4375	11.113
1/2	0.50	12.700
9/16	0.5625	0.5625
5/8	0.625	15.875
11/16	0.6875	17.462
3/4	0.750	19.050
13/16	0.8125	20.637
7/8	0.875	22.225
15/16	0.9375	23.812
1	1.00	25.400

Water transfers heat 3,500 times better than air.

Feet Head to Water Pressure			
Feet Head	Pounds Per Sq. Inch	Feet Head	Pounds Per Sq. Inch
1	.43	100	43.31
2	.87	110	47.64
3	1.30	120	51.97
4	1.73	130	56.30
5	2.17	140	60.63
6	2.60	150	64.96
7	3.03	160	69.29
8	3.46	170	73.63
9	3.90	180	77.96
10	4.33	200	86.62
15	6.50	250	108.27
20	8.66	300	129.93
25	10.83	350	151.58
30	12.99	400	173.24
40	17.32	500	216.55
50	21.65	600	259.85
60	25.99	700	303.16
70	30.32	800	346.47
80	34.65	900	389.78
90	38.98	1,000	433.00

Testing boiler pressure controls

I like using a small drill like this when adjusting the steam pressure control because its much slower than my 18 volt driver and easier to control

Linkage-less burners are 5-6% more efficient than burners with linkages, according to Honeywell.

Boiler Enemies	
Dirt	Can stop valves from closing tight
	Can plug strainers
	Can damage pumps and valves
Air	Inhibits heat transfer
	Impedes flow of steam
Water	Water hammer
	Lowers heat transfer
	Causes premature steam condensation
	Inhibits flow of steam
	Water logs traps
	Could damage valves by wire drawing

Metric

Metric Heat Conversions

1 kcal =	3.968 BTU
1 kgm =	0.00930 BTU
1 KW =	0.948 BTU/sec.
1 kcal/kg =	1.80 BTU/lb.
1 kcal/m^3 =	0.112 BTU/cu. ft
1 kcal/m^{2}h =	0.369 BTU/sq.ft.h
1 kcal/m^{2}h^{0}C =	0.205 BTU/sq.ft.h^{0}F
1 kcal/mh^{0}C =	0.67 BTU/ft.h^{0}F
1 kcal m/h^{0}C m^2 =	8.07 BTU in/sq.ft. g^{0}F
1 kcal/kg^{0}C =	0.999 BTU/lb.^{0}F
1 kcal/m^3 ^{0}C =	0.0624 BTU/cu ft^{0}F
1 BTU =	0.252 kcal
	107.7 kgm
1 BTU/sec. =	1.055 KW
1 BTU/lb. =	0.5556 kcal/kg
1 BTU/cu ft. =	8.900 kcal/m^3
1 BTU/sq.ft.h =	2.71 kcal/m^{2}h
1 BTU/sq.ft.h^{0}F =	4.886 kcal/m^{2}h^{0}C
	1.49 kcal/m h^{0}C
1 BTU in/sq.ft.hr^{0}F =	0.124 kcal/mh^{0}C
1 BTU/lb.^{0}F =	1.001 kcal/kg^{0}C
1 BTU cu. ft^{0}F =	16.2 kcal/m^3 ^{0}C

Metric Pressure

6.8947 kPa	1 pound per sq. in (psi)
9.794 kPa	1m column of water
1 kPa	10.2 cm of water
1.3332 kPa	1 cm column of water
3.3864 kPa	1 inch of mercury (in Hg)
8 kPa	6 cm of mercury
1kg/cm^2	14.223 lbs./sq. in Psi
1 mm WG	0.002927 in mercury
0.0703kg/cm^2	1 lb./sq. in. Psi
340.39 mm WG	1 in. Mercury
43.9 mm WG	1 ounce/sq. inch
6.9 bar	1 psig
0.45 kg	1 pound

Metric Liquid

Metric	U.S.
3.7854 L	1 Gallon
0.946 L	1 Quart
0.473 L	1 Pint
1 L	0.264 Gallons
1 L	33.814 Ounces
29.576 ml	1 Fluid Ounce
236.584 ml	1 Cup

Metric Conversions

KJ/Hr =	Btu/h x 1.055
CMM =	CFM x 0.02832
LPM =	GPM x 3.785
Kj/Lb.=	Btu/Lb x 2.326
Meters =	Feet x 0.3048
Sq. Meters =	Sq. Feet x 0.0929
Cu. Meters =	Cu. Feet x 0.02832
Kg =	Pounds x 0.4536
Kg/Cu. Meter =	Pounds. Cu Feet x 16.017
Cu. Meters/ Kg =	Cu. Ft/ Pound x 0.0624

Convert Temperature Readings

Fahrenheit to Celsius	Celsius to Fahrenheit
$\dfrac{\text{Degrees F} - 32}{1.8}$	$(1.8 * \text{Degrees C}) + 32$

Fahrenheit to Celsius Temperatures

^{0}F	^{0}C	^{0}F	^{0}C	^{0}F	^{0}C
0^0	-17.8^0	11^0	-11.7^0	40^0	4.4^0
1^0	-17.2^0	12^0	-11.1^0	50^0	10.0^0
2^0	-16.7^0	13^0	-10.6^0	60^0	15.6^0
3^0	-16.1^0	14^0	-10.0^0	70^0	21.1^0
4^0	-15.6^0	15^0	-9.4^0	80^0	26.^{0}
5^0	-15.0^0	16^0	-8.9^0	90^0	32.2^0
6^0	-14.4^0	17^0	-8.3^0	100^0	37.8^0
7^0	-13.9^0	18^0	-7.8^0	150^0	65.6^0
8^0	-13.3^0	19^0	-7.2^0	200^0	93.3^0
9^0	-12.8^0	20^0	-6.7^0	212^0	100.0^0
10^0	-12.2^0	32^0	0.0^0	215^0	101.6.^{0}

Definitions:

Air Change: The amount of air required to completely replace the air in the boiler and associated flue passages.

Air, Primary: Air mixed with the fuel to provide combustion.

Air, Secondary: Air mixed with the flue gases to provide proper turbulence to allow complete combustion.

Air shutter: A device to control the airflow to the burner.

Air, Tertiary: Air from the boiler room introduced to the flue to overcome excessive chimney draft. It is sometimes called Dilution Air.

Air to Fuel Ratio: This is the amount of air used in the combustion process. It is typically 15 parts of air for each part of natural gas.

Barometric Damper: A damper installed in the flue piping to control the excessive draft in a Category1 type boiler by introducing boiler room air.

Boiler: A closed vessel to heat water or create steam.

Boiler Feed System: This is a tank and pump(s) which introduce condensate back into the boiler. The pump is controlled by a boiler level control.

Boiler, High Pressure: A boiler, which generates steam pressures above 15 Psig.

Boiler, Low Pressure: A boiler, which generates steam to pressures below 15 Psig.

Boiler, Cast Iron: A boiler, which uses cast iron as its heat exchanger.

Boiler, Steel: A boiler, which uses steel as its heat exchanger.

Boiler, Fire Tube: A boiler where the flue gases travel through the tubes.

Boiler, Water Tube: A boiler where water flows through the tubes.

Boiler, Modular: A heating system consisting of several smaller boilers.

Breeching: A conduit to transport the combustion by products from the boiler to the outside or to the chimney. It is also called a flue.

Btu (British Thermal Unit): The amount of heat required to raise one pound of water, one-degree F.

Btuh: Btus in one hour

Burner: A mechanical device which mixes air and fuel to provide ignition and combustion of the fuel.

Burner, Atmospheric: A burner that uses natural draft and gas pressure to provide combustion.

Burner, Power: A burner that uses an internal blower to mix the fuel and the air for combustion.

Carbon Dioxide / CO2: This is a gas produced as a by-product of combustion or steam.

Carbon Monoxide /CO: This deadly gas is odorless and tasteless. It's produced when the combustion is out of adjustment.

Carryover: This is when boiler water is pulled out of the boiler with the steam. It lowers the efficiency of the system.

Chemical Feed Pump This pump is used to inject water treatment chemicals into the boiler or steam system.

Combustion Air: The air introduced from the outside required for the proper combustion of the fuel.

Combustion Analyzer: A device which measures the flue gas and efficiency of the boiler.

Condensate: Condensed water because of the removal of latent heat from steam.

Condensate Pump: Pump used to return water to the boiler or boiler feed tank.

Condensate Tank: This is a condensate collection tank used to transfer condensate around the building. The pump is controlled by an internal float.

Control, Operating: A device that sense steam pressure & starts or stops the burner. This is usually set for a lower steam pressure than the Limit Control.

Control, Limit: A device that starts or stops the burner. This is usually set for a higher pressure than the operating control. In most applications, it has a manual reset feature.

Cooling Leg: Uninsulated pipe before a thermostatic steam trap that allows the steam to cool and condense.

Counter-Flow System: A pipe where steam and condensate flow in opposite direction.

Dew Point Temperature: The flue gas temperature which is cooled enough to allow the water vapor to condense into water.

Differential Pressure: The pressure difference across the inlet and outlet, typically for steam traps.

Dirt Leg: Nipples and a pipe cap installed just before the train to capture any dirt in the gas line before it enters the gas train.

Draft: The pressure differential between atmospheric pressure and the pressure in the flue and boiler.

Draft Diverter: A device above the boiler where tertiary air is introduced to the flue after the main combustion.

Draft, Mechanical: The pressure differential between atmospheric pressure and the pressure in the flue and boiler that is induced because of a fan or blower.

Draft, Natural: The pressure differential between atmospheric pressure and the pressure in the flue and boiler without a fan or blower.

Dry Return: Return piping above the boiler water line that carries condensate back from the system.

Emergency Door Switch: This switch is installed just inside or outside the boiler room exits. When pushed, it will shut off all the boilers.

Equalizer: A pipe connected between the boiler steam header and the boiler bottom. It equalizes the pressure in the boiler, so the water is not pushed out of the return pipe.

EDR Equivalent Direct Radiation: The amount of heating surface which will produce 240 BTUH. It is based on steam at 215^0F and a room temperature of 70^0F.

Firing Rate: The burning rate of fuel and air in the burner.

Firing Rate Control: A control that senses the boiler steam pressure. It will regulate the burner between low and high fire to meet the desired set point. It's also called the modulating control.

Flash Steam: Condensate above the boiling temperature which flashes to steam.

Float & Thermostatic Trap /F&T Trap: A mechanical steam trap that vents air and other gases through the thermostatic portion and condensate through the float version.

Flue: A conduit that transports the combustion by-products from the boiler to the outside or to the chimney, also called breeching.

Flue Gases: The by-products of combustion produced by the burner & measured in boiler flue.

Fuel Train: The gas pressure regulator, nipples, and gas valves, located in the gas piping directly attached to the burner, also called a gas train.

Gas Pressure Regulator: A device that controls the gas pressure supplied to the burner.

Gas Pressure Switch: A safety device that senses the available gas pressure and will shut the boiler off in the event the pressure is outside of the setting.

Hartford Loop: The Hartford Loop is where the returning condensate is introduced to the boiler. It connects on the equalizer about two inches below the boiler water line.

Heat, Latent: The amount of heat required to cause a change of state.

Heat, Sensible: The amount of heat required to cause a change in temperature.

Heating Medium: The material that the boiler heats. It could be steam, water, or some other type of fluid.

High Fire: The highest design firing rate of the burner.

Lag Boiler: The boiler is not the first boiler to start when there is a call for heat.

Lead Boiler: The boiler is the first boiler to start on a call for heat.

Lockout: A safety shutdown that requires a manual reset of the control or safety device.

Low Fire: This is the lowest design firing rate of the burner.

Low Fire Start: This switch verifies that the burner is in the "Low Fire" position before opening fuel valves.

Low High Low Fire: A burner that starts at low fire and will travel between low & high while there is a call for steam.

Low High Off Fire: A burner that starts at low fire and goes to high fire. The burner will stay at high fire until the call for heat ends.

Low Water Cutoff, Primary: A device that senses the water level inside the boiler and will shut down the burner if the water level drops to an unsafe level. It may also control the feedwater valves or boiler feed pumps.

Low Water Cutoff, Auxiliary: This shuts off the burner if the water levels drops below its setpoint. This one has a manual reset switch, and the boiler will not restart until the water level is restored and the reset button is pushed.

Main Air Vent: This vent is located toward the end of the steam main feeding the building. These vent air so steam can get to the radiators. The vents should be located 15 inches back from the end of the horizontal piping.

Master Trap: This trap is used just upstream of the condensate or boiler feed tank. They can cause water hammer, uncomfortable areas, high fuel costs and should not be used.

Mechanical Return System: These systems use a pump to transfer condensate. They are available in either a condensate or boiler feed tank.

Mechanical Seal: This is between a condensate pump and condensate pump motor. It is used to keep the water in the system.

Modulating Burner: A burner that will operate at any position from low to high fire to meet the demands of the modulating control.

Modulating Control: A pressure control that senses the steam pressure. It will send a signal to the burner that will set the burner at any position from low to high fire.

Near Boiler Piping: This is the piping directly connected to the steam boiler. It is used to dry the steam before going to the system.

Non-Condensing Boiler: A boiler with flue gas temperature above the dew point temperature.

One Pipe System: This heating system has steam and condensate flowing in the same pipe.

Parallel Flow System: This system is where the steam and condensate flow in the same direction.

Pigtail: The curly pipe between the pressure control and the boiler. It allows steam to condense and uses the water to protect the control from the effects of steam, also called a Siphon.

Pilot, Continuous: It is a pilot flame that burns all the time, regardless of whether the burner is firing.

Pilot, Intermittent. It is a pilot that lights when there is a call for heat. The pilot will stay light during the entire time the main burner is firing.

Pilot, Interrupted: It is a pilot that lights when there is a call for heat. The pilot will shut off once the main flame is established.

Pipe Pitch: The amount of slope used to avoid water pooling. It is typically 1 inch slope away from the boiler every ten feet.

Pop Safety Valve: A valve located on a boiler that will relieve the internal boiler pressure if the pressure rises to the rating of the valve. They are sometimes called a Safety Relief Valve.

PPM: Part per Million.

Prepurge: On a call for heat, the burner blower starts to purge the boiler combustion chamber and flue passages of any unburnt fuels. It will operate long enough to provide several air changes inside the boiler.

Pressure Control, Operating: This pressure control will be used to cycle the boiler on and off.

Pressure Control, Limit: This pressure control is a manual reset control, set at a higher pressure than the operating control. The boiler will not restart until the boiler pressure drops below the setpoint of the control and the manual reset button is pushed.

Radiator Air Vent: These small air vents are used on one pipe steam system radiators. It allows the air to leave and return to the radiator.

Safety Relief Valve: A valve located on a boiler that will relieve the internal boiler pressure if the pressure rises to the valve's setpoint. They are sometimes called a Pop Safety Valve.

Sidewall Venting: Boiler flue that is piped to the side wall of the building rather than a chimney or stack.

Siphon This is the curly pipe between the pressure controls and the boiler. See Pigtail.

Spill Switch: A device located by a draft diverter or a barometric damper that senses rollout of the flue gases and shuts off the burner. It is a manual reset switch.

Static Head: This is the weight at the bottom of a column of pipe. It is important when sizing a steam trap or the "A" or "B" dimension on the gravity return system.

Strainer: This is installed upstream of steam traps, condensate tanks, control valves, or pressure reducing valves. It has an internal screen used to prevent dirt, rust, and mud from entering.

Supply Main: The horizontal pipe that carries steam from the boiler to the radiator takeoffs.

Supply Riser: The vertical piping from the boiler to the horizontal steam main.

Swing Joint: The piping used to absorb the expansion & contraction differences between the boiler & the piping.

Total Dissolved Solids: These are primarily inorganic substances found in in water such as salts, magnesium, calcium, potassium, sodium, nitrates, chlorides, & sulfites. These are often referred to as TDS.

Thermostatic Steam Trap: This trap uses a bellows inside which expands when steam is present and contracts when steam is not present. It uses the same orifice to vent air, non-condensable gases, and condensate.

Two Pipe System: This system has a separate pipe for steam and for condensate. They typically have a steam trap on each radiator. Some old two pipe systems were piped into the wet return and no traps were used.

Water Seal, Pressure Controls: These are sometimes used in place of a pigtail or siphon. They provide a water barrier to protect the steam control from the effects of steam.

Water Hammer: This is when steam inside a pipe encounters water. The water causes the steam to collapse rapidly and make a loud banging sound, like a metal hammer hitting the pipe.

WC: This stands for Water Column

Wet Return: This is the condensate return piping below the boiler water line.